KNOWLEDGE AND REFERENCE IN
EMPIRICAL SCIENCE

Knowledge and Reference in Empirical Science is a fascinating study of the bounds between science and language: in what sense, and of what, does science provide knowledge? Is it to be taken literally? Is science an instrument only distantly related to what's real? Can the language of science be used to adequately describe the truth?

Jody Azzouni approaches these questions through a compelling analysis of the "reference" of kind terms. The book begins by investigating the technology of science – the actual forging and exploiting of causal links, between us and what we endeavor to refer to and know. It shows how this technology allows for knowledge gathering, why scientific lore must be regarded as true and in what ways instrumental access to our universe is anchored in observation. The second part of the book studies the language of science and shows how this flexible tool enables us to transform our fragmented investigations of the world into a unitary and seamless discourse. Azzouni proposes that scientific language does allow us to speak of what's true and so he questions influential views which take scientific laws to be at best only approximate, or even false.

Central to the book is the author's support of semantic naturalism – a program of showing that the resources of our language do not conflict with the scientific picture of us as animals. This not only illuminates concerns in the philosophy of language, but also sheds new light on the role of holism and deduction in science.

Jody Azzouni is Associate Professor of Philosophy at Tufts University. He is the author of *Metaphysical Myths, Mathematical Practice: The Ontology and Epistemology of the Exact Sciences.*

KNOWLEDGE AND REFERENCE IN EMPIRICAL SCIENCE

Jody Azzouni

Routledge
Taylor & Francis Group

LONDON AND NEW YORK

First published 2000
by Routledge
2 Park Square, Milton Park, Abingdon, Oxon, OX14 4RN

Simultaneously published in the USA and Canada
by Routledge
711 Third Avenue, New York NY 10017

First published in paperback 2004

Routledge is an imprint of the Taylor & Francis Group, an informa business

© 2000, 2004 Jody Azzouni

Typeset in Times by Taylor & Francis Books Ltd

British Library Cataloguing in Publication Data
A catalogue record for this book is available from the British Library

Library of Congress Cataloguing in Publication Data
Azzouni, Jody.
Knowledge and reference in empirical science/Jody Azzouni.
(International Library of Philosophy)
Includes bibliographical references and index.
1. Science–Methodology. 2. Knowledge, Theory of. 3. Science–Philosophy. I.
Title. II. Series.
A175 .A995 2000

501–dc21   99-088208

ISBN 978-0-415-22383-6 (hbk)
ISBN 978-0-415-33354-2 (pbk)

CONTENTS

CONTENTS

PREFACE

When I first conceived of writing a book, back in 1989, I imagined a work spanning both the contents of this book and my previous one in the philosophy of mathematics. The view is a synoptic one, although I soon realized that mathematics raised enough special issues to warrant separate treatment. After my 1994 book went to press, I set about writing this one, and it was substantially complete by January of 1995. I regard it as continuing the kind of epistemic–linguistic analysis that I had started in the first book. For reasons of time, revisions in the summer of 1999 had to be restricted to acknowledgments – where pertinent – of my own publications in the intervening time, and to judicious additional engagement with the literature.

The philosophy of science literature, in particular, is enormous and extremely rich. I very much regret not writing more about it – but given the project this book undertakes, I have to restrict myself to the literature bearing directly on the issues I raise, and which doesn't require too fine an analysis of disagreements.

I especially regret not discussing further the papers that emerged around the time I wrote this, and after, in the self-styled philosophy of chemistry. There is a lot there that I would love to comment on; but this is not the time and place. Similar remarks apply to newer literature in the philosophy of science that endeavors to show how experiment "has a life of its own": I wish I had had the time and space to comment on more of this work than I have done here.

Let me say something about the issue of sub-disciplines. It should be no secret that the major trajectories in philosophy in the twentieth century, at the hands of people like Quine and Putnam, range from philosophy of science and logic – narrowly construed – to philosophy of language. It is views that Quine held about confirmation, and about how scientific doctrine evolves, that molded his perspective about purist issues such as analyticity, meaning, and indeterminacy. So, too, Putnam's early work in the philosophy of science and of mathematics inspired his later thought experiments about reference and meaning. This is not a simple matter of cute analogies: There are important conceptual links here that we ignore at our peril but which the movement towards sub-disciplining in philosophy *has* caused us to ignore. One small example: you would think that philosophy of mathematics and philosophy of science are close enough thematically for fruitful engagement. Nevertheless, the literature on indispensability, and its presuppositions – as found in Quine, Putnam, and subsequent philosophy of mathematics – seems unaffected and unaffecting of philosophy of science views, like those of Cartwright, van Fraassen, Boyd, and others, who quite comfortably take fundamental scientific law to be false or approximate. Both these literatures, in turn, seem detached from the topic of the role of the truth-predicate, as it comes up in the philosophy of logic – at least if you rely on the citation record. And yet, all this stuff is thematically linked in quite intimate ways.

This, I guess, is an apology – in the old-fashioned sense – for the sweep of this book: From topics in the philosophy of science at the beginning to philosophy of language at the end, and the philosophy of logic and of mathematics all along the way. Synopsis must pay off in illumination. Hopefully it does so here.

ACKNOWLEDGMENTS

Thanks are especially due to Arnold Koslow for twice providing detailed comments on this book, once in 1995, on the penultimate version, and again in 1999, when I was finally engaged in last revisions. His suggestions were numerous and thoughtful. I'll probably regret not taking all of them. Copious thanks are also due to George Smith, with whom I've had uncountable conversations on these topics, from the time I started working on the book, and before. Both he and Babak Ashrafi (to whom I'm also very grateful) provided almost line-by-line comments (invariably invaluable ones) on Parts I and II in the summer of 1999. I would also like to thank Mark Richard and Stephen White for several useful suggestions. Thanks also to my Scientific Realism seminar class in the fall of 1994, during which I presented a version of this book, and again to George Smith, who attended several sessions of that class, and to Scott Tanona. Finally, my thanks to several people who provided critical suggestions anonymously, and to others whom I've forgotten during the long interim during which this book incubated. My apologies for not naming them individually.

I also want to thank the City University of New York Graduate Center for inviting me to be a visiting scholar for the academic years 1993–94, and 1996–97, and to Tufts University for graciously giving me the time off to pursue these studies. Lastly, my thanks to the philosophy department at Tufts University, and especially to Julie Roberts and Anne Belinsky, for unfailingly providing an environment in which philosophy can flourish.

GENERAL INTRODUCTION

I pursue the two-headed theme, with regard to empirical science, of what there is and how we know what we know about it.[1] I find these topics linked, and this book is one long intricate argument around which they intertwine. I won't give the argument here – the book itself does that – but I offer a schematic overview to provide the reader a fair sense of where things are going and how they fit together.

Parts I and II are concerned largely with epistemology. Part I attacks two widely accepted doctrines, the first being *confirmation holism*, the doctrine, roughly, that it is not sentences or statements which are confirmed or disconfirmed, but only larger units such as groups of sentences or theories. The second doctrine I call *theoretical deductivism*,[2] which is the view, briefly, that scientific theories, either alone or with other scientific theories, are testable or empirically applicable by direct comparison of derived consequences against experience, and *without the benefit of other intermediaries*. This second doctrine is *very popular*. Positivism had a version of it, according to which the testable consequences are "observation sentences." But the view easily survives confirmation holism, and then the derived consequences are theory-laden: Sometimes one writes as if our entire collective body of beliefs is a single scientific theory tested *in toto* against experience.[3]

---

1   Mathematics and logic pose special cases dealt with in Azzouni (1994). Occasionally I allude to positions taken there, but not in a way that prevents this book from being self-contained.
2   I am grateful to Arnold Koslow for suggesting this name.
3   Quine writes this way sometimes, but then he'll take it back:

> The holism in "Two dogmas" has put many readers off, but I think its fault is one of emphasis. All we really need in the way of holism ... is to appreciate that empirical content is shared by the statements of science in clusters and cannot for the most part be sorted out among them. Practically the relevant cluster is indeed never the whole of science; there is a grading off, and I recognized it, citing the example of the brick houses on Elm Street.
>
> (1980: viii)

1

Philosophers are not alone in their adherence to theoretical deductivism as this statement from a first-year physics textbook indicates:

> A scientific theory usually begins with general statements called *postulates*, which attempt to provide a basis for the theory. From these postulates we can obtain a set of mathematical laws in the form of equations that relate physical variables. Finally, we test the predictions of the equations in the laboratory.[4]

My opposition to confirmation holism is one of *degree*, not principle – the holism is overrated – but it's enough of a disagreement to argue over in print: Certain standard scientific practices are misconstrued because of the presumption that confirmation holism is at work where it simply is not.

Theoretical deductivism, however, is *false*: in testing or applying a theory or group of theories, the derivation of predictions is often infected with additional and crucial statements belonging to no theory or even proto-theory. I call these statements gross regularities, and I show that they are (epistemically speaking) quite independent of scientific theories. The tendency of many philosophers to invoke confirmation holism, coupled with the concomitant elevation of humble but largely unexplained truisms about sensory experience, materials used in experiments, and so on, to genuine scientific dogma, enables those philosophers to overlook the unique role gross regularities play by assimilating them into nebulous empirical theories where they are idealized beyond recognition into entirely articulated scientific theory. Part I illustrates the importance of gross regularities to our understanding of how evidence for scientific claims is gathered; Part II shows how gross regularities provide for the accumulation of scientific knowledge despite incommensurability.

One objection the *deductive holist* (one committed to both confirmation holism and theoretical deductivism) can raise should be dispensed with now: Since gross regularities are part of our web of belief, they are part of the deductive structure we use to apply and test science; only a question-beggingly narrow interpretation of "scientific theory" allows gross regularities an independent life-style outside of scientific theory.

The response is that gross regularities operate in an epistemically distinct way from genuine scientific theory:

1   They must often be empirically established in the experimental or application situation in which the web of belief is itself being tested and applied, and their scope and range must almost always be experimentally delineated in that situation as well.

---

4   Resnick *et al.* (1992: 469).

2   They are independent of scientific theory in the sense that they will survive the collapse of scientific theories, no matter how drastic.

3   Many of them (unarticulated gross regularities) cannot be described as belonging to a web of *belief* without a notion of unarticulated belief in which to house them.

Conclusion: Deductive holism is an epistemic doctrine that obliterates crucial epistemic distinctions *within* our web of belief.

One philosopher of science opposed to something similar is Peter Achinstein. He attacks "hypothetico-deductivism," the doctrine that "the scientist thinks up some hypotheses comprising a theory from which, together with auxiliary assumptions, he or she derives conclusions. The theory is tested by determining the truth of these conclusions" (1991: 317).

Although I like Achinstein's views on experiments (see pp. 35–7), my perspective is very different. This is because Achinstein thinks

> testing a theory experimentally requires (1)' deriving theoretical consequences from it, (2)' establishing the truth of experimental result claims by experiment, and (3)' making *nondeductive inferences* from experimental result claims to the theoretical consequences.
>
> (1991: 321, italics mine)

That is, he thinks *deduction* from theory to experimental situation is often absent.[5] My focus is different: although I agree with his claims, I am interested in the rich number of cases in which generalizations of a certain sort – gross regularities, epistemically independent of science in the sense described above – arise. Strictly speaking, therefore, my view is compatible with the success of hypothetico-deductivism, since gross regularities, however discovered and when, can take their place in deductive inferences. Also there are violations of theoretical deductivism in many more places than Achinstein sees violations of hypothetico-deductivism.

Part II gives a new picture of evidence gathering in empirical science, one strictly opposed to theoretical deductivism. Evidence gathering occurs within a two-tiered coherentist structure where, roughly, the bottom tier provides the framework for evidence that confirms scientific results found in the upper tier. Crucial to lower-tier evidence gathering are *procedures*: methods we use to learn about the world; and it is by means of gross regularities

---

5   Boyd (1985) also stresses the importance of nondeductive methods – induction – especially in respect to explaining experimental artefacts; and one should not overlook Hempel's 1988 presentation of the problems with the inferential construal of scientific theorizing. (My thanks to Arnold Koslow without whom *I* would have overlooked the latter article.)

that we "operate" procedures. Procedures are how we (causally) interact with things in the world, and thus they link the topics covered in Parts I and II, and the issue of how reference and causality are connected, which forms the major subject of Parts III and IV.

A remark on terminology. Some philosophers use *causal* so that what is causally related *must* also be related by scientific law (under, perhaps, a description). I do *not* use the term that way; rather, I use it as do philosophers who speak of the "causal theory of reference." Causes thus understood may well turn out to be epiphenomenal – not genuine constituents of our universe; in any case I take generalizations about them to be gross regularities. I say more about this in Part III.

Parts III and IV. Programs for naturalized epistemology and semantics are popular these days, where such programs aim to handle semantics and epistemology compatibly with our current scientific picture of ourselves.[6] Philosophy, as Quine puts it, is continuous in its methods with empirical science. So, roughly (for different naturalization programs are rampant), epistemology is a provincial topic snug within one's conceptual scheme, illuminated by the pertinent psychological literature and evolutionary considerations.[7] Similarly, one attempts to show that semantic terms like *refer* pick out (or supervene upon) unproblematical causal relations that animals like ourselves can have; and one attempts to show that inference patterns, licensed by rules governing logical connectives, are underwritten by psychologically real processes.[8]

There are opposing voices, of course, raising specters of scepticism, normativity, intrinsic intentionality, and worse, who claim that naturalization cannot be carried out without emptying still-living philosophical issues of their content, and who mutter dark sayings about babies and bathwater.[9]

Consider the naturalization program specifically with regard to reference. Causality is central, for one way or another it is causal relations between the

6   The glossary of Boyd, Gasper and Trout (1991: 778) defines naturalism thus: "The view that all phenomena are subject to natural laws, and/or that the methods of the natural sciences are applicable in every area of inquiry." See Azzouni (forthcoming) for details on different ways to understand "naturalism."

7   See Dretske (1981); Goldman (1986); Kim (1988); Putnam (1983d); Quine (1969a, 1969b); Stroud (1981); see also the articles in Kornblith (1985) and Radnitzky and Bartley (1987), among many others.

8   Examples are, among others, Devitt (1991); Field (1972); Kripke (1980); Putnam (1970, 1975a), and some of the articles in Schwartz (1977). Also see the closely related literature on mental content: Dretske (1981); Field (1981); Fodor (1987, 1990b, 1990c), Stalnaker (1987); Millikan (1984). Both Kripke and Putnam are, at best, ambivalent about naturalism; and neither wishes to establish a "causal *theory* of reference." Recall Kripke's remarks about wanting to offer "a *better picture*" (1980: 93); see Putnam (1992: 221, note 4).

9   See Blackburn (1988); Putnam (1983d); Searle (1983, 1984); Stroud (1981); Wilson (1982), among others.

use of a term and the object(s) referred to by this term that are supposed to provide the naturalization for reference. Philosophers are deadlocked on whether causality can do the job. On one side there are those who think a robust set of causal relations can not be picked out that will ground our apparent talk of reference; such philosophers commonly invoke what I call permutation arguments or the Löwenheim–Skolem theorem.[10] Opposing them are those who think causal relations (with, perhaps, other entirely natural resources, such as descriptions) suffice to naturalize reference.[11]

Permutationists and Löwenheim–Skolemites almost invariably try to show that causality, plus the other resources available, fail to underwrite the referential relations between our empirical terms and the world in a particularly dramatic way: It is not that empirical terms are partially fixed by such resources, and otherwise partially open to reinterpretation;[12] rather, the available resources fail to provide *any* anchoring for such terms. Consistent with causal relations, descriptions, and whatever else the semantic naturalist conjures up to fix reference, empirical terms are still vulnerable to reinterpretation. The upshot, usually, is inscrutability of reference,[13] a linguisticized idealism that applies to all the terms of a language so that one cannot distinguish where the use of a term is naturalistically connected to the objects referred to, and where it isn't so connected; and, indeed, Putnam cites Kant to give philosophers a take on his version of the view.[14]

Part III shows that permutation arguments and the Löwenheim–Skolem theorem don't affect the viability of programs for naturalizing semantics: they don't significantly bear on whether causal relations, or other resources, can fix the reference of empirical terms. The issue has been aired a bit in the literature, and so it may seem that there isn't much new to say here. The

10 Here is an incomplete list of articles: Davidson (1977, 1979); Field (1975); Putnam (1977, 1978b, 1981, 1983e, 1983a, 1983b, 1984, 1986, 1989); Quine (1964, 1969c, 1981c, 1986b); and Wallace (1979). Permutation arguments, of course, are usually applied to *all* terms of a language, not just empirical ones. But I will concern myself only with empirical terms for which the possibility of using causality to help anchor reference is an option.

11 Examples are Devitt (1983, 1991); Field (1972); Fodor (1987, 1990b, 1990c); Glymour (1982); Hacking (1983); and Lewis (1984). I should add that, given his views, Field properly belongs here rather than in my first list. He is there, too, however, because he uses permutation arguments to motivate what he calls "conventionalism." See pp. 160–2

12 One might treat vague terms this way: Our sensory capacities partially fix the extension and anti-extension of terms like "red," leaving an indeterminately vague border. The semantic naturalist can accept this sort of indeterminacy, but, with the exception of Field (1975), the philosophers I am speaking of have something far more extreme in mind.

13 The term is used in Quine (1969c). When I say empirical terms are open to reinterpretation, I don't mean that the reinterpretations available for each term are independent of the reinterpretations of other terms. On the contrary.

14 Putnam (1978a: 5–6). Quine, too (1986a: 492), calls his *ontological relativity* a "rather Kantian thesis."

topic is not dead, however, in part because realist attacks on permutational arguments have *not* been fair ones: referential realists too quickly invoke a simple use/mention confusion between causality and "causality."[15]

Unfortunately, clearing away the above won't do as much for semantic naturalism as its proponents may have hoped, for there are still two tasks remaining. The first is to show that the referential relations apparently required by our semantic practices can be naturalized (say, by underwriting them causally). There are reasons to think this false, even apart from the concerns permutation arguments raise: although the resources available to naturalists can partially underwrite the referential relations we take our terms to make to things in the world, they fall short in crucial ways.

But the more important and primary task facing semantic naturalists is, I think, this: there is something deeply peculiar, *philosophically* peculiar, about the program of naturalized semantics on view. One wants to know what it is about our referential practices that *requires* referential relations to be underwritten by, say, causal relations; and, surprisingly, proponents of such programs have almost nothing to say about this.[16] Perhaps this is because it seems obvious that a commitment to naturalism requires some sort of *mechanism* to explain how our terms refer to what they refer to. But in fact it is not obvious at all that a mechanism is needed; and, after the matter is carefully investigated, it will be clear that neither our referential practices nor a commitment to naturalism support these projects so beloved by semantic naturalists.

The above explains puzzles about reference previously noticed by philosophers: its normativity, and its peculiar immunity to certain kinds of mishap. These are symptoms that causal relations, although playing a significant role in how we use terms, do not play quite the role the causal theorist envisages for them. Language in the empirical sciences and in ordinary life intricately combine (1) a body of verbal practices surrounding *truth* and *reference* which presuppose reference to be a settled and fixed relation between us and (sets of) objects, and (2) the actual forging of robust (in the sense that gross regularities are robust), but gerrymandered (partial, parochial, and revisable), causal connections between empirical terms and what they refer to.

The resulting picture not only explains why we have the semantic practices we have but also soothes the genuine philosophical perplexity about reference that motivates so much of the literature in this area.

---

15 Devitt (1983) and Glymour (1982) independently direct this accusation towards Putnam (1977) and (1978b). Field (1989: 277 note 44), and Lewis (1984) also attribute the error to Putnam, and I have the impression they are not alone. In fact, as Part III shows, this objection is pretty much baseless.

Sosa (1993: 606) writes: "It seems to me that on the [model-theoretic] argument we have reached an impasse." I hope to change this.

16 Putnam (1981: 1–5) argues, in a semi-popular exposition of the causal theory of reference, that reference does not take place by "magic"; but this is hardly adequate.

The reader may fear, despite my protestations, that my offer is not friendly to naturalism. On the contrary (I protest again): the situation is somewhat analogous to one I've discussed elsewhere, namely, the ontological status of mathematical posits and its bearing on the naturalist program. It is worth examining this case briefly to illustrate my approach to ontology and show how it is compatible with naturalism. What follows is a severe abbreviation of Azzouni (1994).

Here is the problem. Mathematical practice seems to commit its advocates to an ontology of abstract objects – numbers, functions, sets, Banach spaces – objects which don't interact with us, or with the rest of our world. Such an ontology seems incompatible with naturalized programs in semantics and epistemology which presuppose an interaction with the objects we refer to and know about. I call this the *naturalization problem*.

The naturalization problem has convinced most philosophers of mathematics that they have a hard choice: *either* accept naturalized epistemic and referential constraints, and tamper with either the ontology or the truth-value status of mathematical language, *or* deny the naturalization program with regard to epistemology and semantics – adopt, that is, a version of *non-naturalism*. One drawback to strategies compatible with the first choice is that they do not seem true to mathematical practice: mathematical language is identical in its workings to the operation of language elsewhere. "There is one even prime," we say cheerily, just as we say, "There is one Queen of England." And in the same way we attribute truth and falsehood to statements about objects, too: "The even prime is greater than seventeen," we think false in just the same way we think "The Queen of England has multiple personality syndrome," false.[17] Thus any philosophical program to rewrite mathematical language and/or its *prima facie* semantics, in order to eliminate the ontological and truth commitments apparently located there, seems misguided. The set of approaches compatible with the second choice are unacceptable because they, one way or another, leave unexplained *how* we have the mathematical knowledge we have, and can refer to the mathematical objects we can refer to.

There is another way: naturalization programs need not run so directly towards their goal. Consider talk of reference and the natural mechanisms we have for explaining how we refer to what we refer to, e.g. one or another sort of causality. Naturalization proponents think their programs must supply mechanisms that fit perfectly with the talk of reference and truth they underwrite. It is this presumption that usually lies behind the strategies

---

17 That there are *two* aspects to this similarity needs stressing. First, T-sentences such as " 'Snow is white' is true *iff* snow is white" are taken to hold for mathematical sentences, as for other sentences. Second, mathematical statements are taken as true or false just as are statements in other domains of discourse. *Linguistic realism* is the commitment to *both* these claims with regard to a domain of discourse.

compatible with the first choice. But what in the naturalization program requires this? Consider, instead, a collection of *posit-terms* with the following properties:

(a) We apply talk of truth and talk of reference to such terms in a way identical to how such talk is applied to other "genuinely referring" terms.
(b) We supply a collection of (possibly revisable) procedures for recognizing true sentences containing such posit-terms.
(c) We disallow any requirement to explain how we are able to refer to such terms. ("They are *posits*," we say testily, and mean that no such requirement is necessary. There is a sense, after all, in which posits are "made up.")
(d) We explain how we know what we know about the "objects" the posit-terms refer to by pointing to the procedures for recognizing true sentences given in (b), and note that no further explanation – of how such procedures provide knowledge of the objects – is needed. (Again, we say, rather testily, "They are *posits*, and so the procedures are *stipulated*.")

Illustration: Start with a *material-object* language, and presume a causal story about how we refer to, and know about, material objects; that is, tell a (causal) story about how we refer to chairs by means of the term "chair," and supplement this with a reliabilist (or whatever) account to explain how we know what we know about chairs. Then, assume a normal (Tarskian, say) truth predicate for this language so we can say " 'Chairs are fun to sit on' is true *iff* 'Chairs are fun to sit on,' 'Chairs are designed to be sat on' is true, and 'Chair' refers to chairs."

Now augment this language with terms from second-order Peano arithmetic (hereafter SOPA), introduce axioms for SOPA along with a (necessarily incomplete) set of axioms and inference rules for second-order logic, and *stipulate* that any sentence S in the language of SOPA is true *iff* a derivation of S from these axioms exists. Extend the truth predicate and the referential apparatus of the material-object language to the language of SOPA; and, six days having passed, rest.

Notice that we've solved the naturalization problem. No hard work in naturalized epistemology or semantics is needed since numbers on this view are *posits* (no story is needed about how we gain epistemic or referential access to them), and the procedures we use to learn the truths we learn about them are *stipulated* (no story need be told about what justifies the methods we use to learn about numbers). This should give mathematical realists (some) reasons for cheering, for no rewriting of mathematical idiom is required: The semantics and talk of truth smoothly apply over both the material-object language and the language of SOPA, and the language of SOPA is taken at face value.

A possibly awkward question remains: do mathematical objects *exist* on this picture? Some think condition (c) implies that they don't. Others – those

Quinean about ontological commitment – will claim that by adopting (a) we've committed ourselves to the existence to numbers (and probably of sets, too).

I'm not sure that this question matters (or entirely makes sense, at least in the broad way usually posed).[18] In any case, this is the crucial point: an explanation has been given of how we know about and refer to numbers in a way the naturalization proponent should not object to, and which the realist, insofar as her motives arise from a desire to leave mathematical talk as is, should be equally satisfied with.

It's fairly transparent that the literal story just given about SOPA *won't* do as it stands. For one thing, it's hard to stipulate (given the usual interpretation of Gödel's theorem) that all and only the truths of arithmetic are *deducible* from these axioms. Issues about the justification of applied mathematics are also not addressed by my little parable.

Happily, we can tinker with this story so that the result is adequate to mathematics without compromising its virtues vis-à-vis naturalization.[19] What remains invariant after introducing the requisite changes is that mathematical objects are posits, and that what is true of them is stipulated to be so true.

What moral follows from this example? This, and I'll illustrate it again in Parts III and IV of this book: naturalization does not require, when it comes to reference, an actual reduction of the referential relation to other (natural) relations, or even a showing that the referential relation supervenes on such natural relations. For there is another way: we can explain why we treat reference the way we do (why we use the theory of truth we use, why we take our terms to refer the way we do), and this explanation can be through-and-through compatible with naturalism *even if the referential relations that supposedly exist between us and items in the world aren't.*

18 Reasons for taking this position can be found in Azzouni (1998).
19 See Azzouni (1994).

Part I

PROCEDURAL FOUNDATIONALISM

The final arbitrator in philosophy is not how we think but what we do.

(Ian Hacking)

§ 1

INTRODUCTION TO PART I

Once upon a time epistemology was *soothing*. To stay out of trouble it counseled reliance on undoubtable sense experience, and inference from that basis by *a priori* truth. Testing beliefs, therefore, was easy. You deduced observational consequences from them, and then observed if they were true.

Quine inherited from Carnap a diminished version of this view, and repudiated it. No foundation for our knowledge claims is possible, he argued, since no purely observational language is possible. The commonsense conceptual scheme in which we live and breathe is impure, each precious concept, however observational, tainted by theory. We may replace the scheme with something better, and, indeed, replace our native tongue with a formal prosthesis; but any synthetic product, whatever its virtues, will not cleanly divide the unassailable observation from the dubious inference. As for insight into incorrigible mathematical and logical truth, holism dictates that in surrendering to the demands of recalcitrant experience, any purported truth, no matter how treasured, may end its day on the chopping block.

The result is that testing your beliefs is not as easy as the positivist view made it seem: When you look around to see if your "observational" consequences are true, you are, in principle, testing your *entire* set of beliefs. The flip-side of the claim, one paragraph above, that any purported truth can go, is that any purported truth can stay (regardless of how the world intrudes), provided we tinker with other bits of what we believe. Our methods of confirmation and infirmation, thus, are tainted by vague methodological canons such as simplicity and conservation.

Quine has company now: due in part to his influence, these views have become clichés of contemporary epistemology. Parts I and II, therefore, challenge this picture and substitute another. In Part I, I revive a form of epistemic foundationalism, although not the linguistic form crushed so thoroughly by Quinean considerations. Rather, I shift from attempts to separate linguistic components of our conceptual scheme which are (more or less) observational from those which are (more or less) theoretical, and attend instead to the methods used to connect our terms to the things we refer to.

13

Here's the script for Part I. §§ 2 and 3 attack confirmation holism and theoretical deductivism, at least in the forms they take at the hands of recent philosophers of science. I show that the impact of confirmation holism on the epistemology of science is far less than philosophers make of it. In particular, confirmation holism is muted across the traditional fault lines between sciences; indeed, these fault lines are not merely ones of subject matter: they are *epistemic* fault lines.[1] I next mount an attack on theoretical deductivism, and then in § 4 discuss what the autonomy of the special sciences comes to. § 5 examines the role of the truth predicate in our web of belief and uses the results against Nancy Cartwright's brand of scientific anti-realism. In § 6, I turn attention to a class of regularities (what I call "gross regularities") which are the soil from which the evidential procedures for the sciences spring. It is the peculiar fact that some gross regularities must simultaneously function as an evidential basis for the sciences while remaining (for the most part) impermeable to scientific theory which gives procedural foundationalism its unusual character.

At this point I'm positioned to discuss the methods, mentioned one paragraph back, used to connect our terms to the world. These methods, called *procedures*, are ways of forging and exploiting causal connections between ourselves and what we talk about: their operations are licensed largely by gross regularities, and as a result they enjoy partial epistemic independence from empirical science, too. This is the central tenet of *procedural foundationalism*.[2]

A further claim (§ 7) is that certain special *(perceptual) procedures* are foundational to all our causal practices. I mean by this, first, that such procedures are in a strong sense (how strong, I make clear in § 8) not replaceable, and, second, that they are (*literally*) proximate parts of *all* the procedures we use to connect our terms to the world.

Procedures should be studied carefully by those interested in the "causal theory of reference." But my concern in Part I is not so much with what empirical terms refer to as with the general epistemic properties of the methods we use to get at what such terms refer to. The *term* will not be neglected, however: I turn to it in Part II.

---

1   Contrast this with Fodor (1990a: 183); I also recommend the reader take a peek at Hacking's discussion (1990: 157) of Boutroux. Although I am not entirely unsympathetic to those who see metaphysical reasons for the existence of special sciences, I intend to concentrate much more, although not entirely, on the (generally underrated) epistemic reasons for such sciences.

2   I discuss procedures in pretty fair detail in what is to follow: for now, for concreteness, the reader may imagine seeing amoebas through a microscope. The procedures, in this case, are the methods the viewer uses via the microscope to learn about the properties of amoebas (I will not be very precise about how to individuate procedures); the gross regularities are, among others, various hands-on tricks a viewer uses to distinguish, for example, artefacts of the microscopic procedures from actual properties of amoebas.

These days, *meaning* holism,[3] as it's called, gets a lot of attention because of its apparent impact on central doctrines in linguistics and the philosophy of mind. Almost everyone who distinguishes between confirmation holism and meaning holism unconcernedly concedes the former to Quine, while going on to worry about what argument can enable a deduction from that (uncontroversial) holism to the interesting (if possibly incoherent) claim of meaning holism. My concern, by lonely contrast, is not with meaning holism at all: I investigate the scope of confirmation holism, for that holism is the one bearing on the epistemological issues involved here.

Despite my adoption of procedural foundationalism, I call the general epistemic position developed in Part II "two-tiered *coherentism*." That is, the epistemology for sentences (the methods by which sentences are confirmed and disconfirmed) is coherentist in contrast to that for procedures.

Here's why. Procedural foundationalism recognizes that the causal mechanisms used to connect ourselves to the world have a certain (epistemic) rigidity. But this does not require *terms* to have fixed sets of causal mechanisms ("operations") associated with *them*, as operationalists hoped.[4] *Vocabulary* can change a lot even if the epistemic engineering embodied in our causal resources remains untouched.

A corollary is that procedural foundationalism has only indirect impact on the plasticity of conceptual change. Although the broad sorts of claims one finds in, say, Churchland (1979), Feyerabend (1962), Kuhn (1970), Quine (1951), etc., *cannot* be sustained, the argument is complex, as I'll show in Part II, and doesn't require rigid connections between sets of perceptual procedures and "observation terms."

I conclude with an important methodological point illustrated in Parts I and II: our view of how we gather evidence and do science is seriously distorted if we concentrate only on how and whether vocabulary changes, and not on how and whether our evidential procedures do.

---

3   The doctrine, roughly, that sentences or statements are not meaningful on their own, but only in a larger context such as a group of sentences, a theory, or an entire language.
4   Part IV shows how the presence of this slippage, between terms and the procedures used to pick out what such terms refer to, creates insurmountable difficulties for causal theorists, such as Fodor, who hope for nomological connections between our use of empirical terms and what they refer to.

§2

PROGRAM AND SCOPE

By a *program* for a science I mean a (sometimes vague) description of what its terms hold of, and on which its laws are supposed to operate. In practice, what falls under a program turns, broadly speaking, on a (fairly unified) set of methods available to practitioners, and on certain unities in the domain the science is applied to (which explain why the methods work where they do). Ultimately, however, the *possible* range of these methods (however refined) is taken to be underwritten by physical fact. Thus, the program of physics is widely believed to be "full coverage," as Quine puts it – *everything* is supposed to fall under its sway;[1] chemistry studies the composition, structure, properties, and reactions of matter insofar as this turns on what we now know to be electromagnetic forces; organic chemistry studies the compounds of carbon; biology studies living organisms and life processes; etc.[2]

---

1 Presumably *abstracta* don't. Whether, therefore, the program of physics *is* full coverage seems to turn on whether *abstracta*-realism is right or not. This issue is terminological, however, since the program of physics can be *full coverage* with respect to the other empirical sciences.

    Strictly speaking, it is often *theories within* a subject matter that have programs and scopes rather than the subjects themselves. For example, *mechanics* was for a long time taken to have full coverage as its program, and the decline of that program was due to developments within physics itself (e.g. Maxwell's electromagnetic-field theory). However, it is an interesting fact that, historically speaking, one or another physical theory is often taken to have full coverage for its program: Nagel (1961: 288) notes that once mechanical explanations of electromagnetic phenomena were abandoned, serious attempts to treat mechanics as a special case of electromagnetic-field theory arose. This sort of historical phenomenon supports the Quinean claim that it is physics, as a whole, which has full coverage as its program rather than some theory within it.

2 Since our assumptions about kinds, their properties, and what principles they obey can be so wrong, there can be slippage in programs. The alchemical search, for what was essentially a chemical method of transforming certain substances into gold, relied on the mistaken assumption that gold is amenable to chemical manipulation the way, e.g., lead oxide is. (In making this point I do not assume alchemists had anything like the current notion of *compound* or of *element*.)

Scope is the domain of application of the laws and techniques of the science achieved *to date*. It depends less on having the right laws for a science than it does on technical and mathematical tools available which mark out how far the laws in question can be tested and applied.[3]

A brief digression. My subsequent concern is with the link between intractability (that is, the factors blocking the direct application of laws and truths of a science to areas we would otherwise expect them to apply) and the methodological independence of special sciences. I concentrate, therefore, on how scope limitations give rise to special sciences. But of course there would not *be any* special sciences if there were no domains obeying general regularities susceptible to recognition in a direct way without needing a reduction to something more fundamental. Here purely physical considerations, not epistemological ones, provide an explanation of what is going on. Programs in special sciences (and in physics, too) are ultimately to be explained physically.

Consider chemistry first. The reason chemical processes obey laws that can be studied somewhat independently of physics is that nuclear forces decay many magnitudes faster than electromagnetic ones. Chemical processes operate by means of atomic bonding via electron exchanges of various sorts; and the properties of elements, and the chemicals made from them, can be explained by generalizations relying on these exchanges alone.

Something similar may be said about biology. Regularities arise here because there are self-replicating (modulo factors which introduce variations) organisms which respond selectively to external pressures. Therefore we can explain why such regularities (and laws) arise without necessarily being able to reduce them to laws or regularities of the underlying sciences; we can, for example, explain why teleological explanations arise in biological contexts.[4] The relation between psychology and neurophysiology is similar. *End of digression.*

The program–scope gap: illustrations from physics

I now illustrate the program–scope gap in empirical science indirectly by showing how *successful* confirmation of scientific theory, and *successful* application of such theory to special contexts, indicate this gap. I discuss five approaches: the development and application of sophisticated mathematics;

---

3 "Domain of application" should not be understood as that domain wherein the laws and techniques of a science have *actually been* applied to date (*these* particular planets, say, although not *that* one because it happens that no scientist has gotten around to studying it), but rather more broadly, to that domain where it seems clear and unproblematical how to apply the laws and techniques we've developed; "clear and unproblematical," largely because we've handled these sorts of cases before.

4 We can also explain why chemical kinds are crisply individuated, and biological kinds are not: electrons carry charge in discrete units, whereas species are made up of organisms which respond differently to selective pressures.

the use of calculational short-cuts; the deployment of idealized models; the design of laboratory experiments; and the use of formal heuristics.

The development and application of sophisticated mathematics

An enormous amount of eighteenth- and nineteenth-century mathematics arose from the need for methods of solving differential equations, and, where solutions are not to be had, methods of deriving theorems which shed light on the motions described by such equations.[5]

Differential equations arise in all branches of physics: celestial mechanics; the equilibrium of rigid bodies; elasticity; vibrating strings and membranes; pendula; heat; fluid dynamics; electricity; electromagnetic theory; relativistic and nonrelativistic quantum mechanics; etc.[6] Such equations can (but don't *always*) characterize what they describe deterministically. For example, consider the Newtonian *n*-body problem. Initial conditions plus the inverse-square law completely determine the locations, velocities, and accelerations of *n* point-masses for all time. This describes the *program* of (this bit of) Newtonian mechanics. Its scope, however, is limited by, among other things, our capacity to mathematically manipulate the differential equations in question to extract the physical information we need.[7]

The absence of general methods for solving differential equations leads naturally to an explosion in mathematics: Numerous techniques must be developed for solving specific classes of such equations. Such techniques are often quite restricted in their application, and offer little insight into why they apply where they do, and what (if anything) can used instead where they don't. Since broad and comprehensive principles for the solution of differential equations are lacking, the result is a morass of separate techniques for the various types of equation.[8]

---

5 See Kline (1972), especially Chapters 21, 22, 28 and 29. For a taste of the subject itself, see Courant and Hilbert (1937) and (1962).

6 Differential equations do not arise just in physics, of course. They occur wherever dynamical processes and something like force laws are available, e.g. in evolutionary theory or in connectionist models for cognitive processing. On the latter, see some of the articles in Rumelhart *et al.* (1986), especially Smolensky (1986). For neat elementary applications of differential equations to biology and sociology, see Braun (1993).

7 For a nice discussion of the *n*-body problem, see Suppes (1984: 125–30).

8 An analogous example is the study of decidable and undecidable subclasses of the predicate calculus. Here, too, complex cases must be tackled by limited and specialized tools that offer little general insight into what, if anything, solvable subclasses of formulas have in common. The rich (although frustrating) resulting subject matter indicates graphically the epistemic sacrifice (i.e. the undecidability of the predicate calculus) originating it. See, for details, Dreben and Goldfarb (1979) and Lewis (1979). Of course, this example is restricted, pretty much, to pure mathematics: Since the subject is so little applied, it does not illustrate how the scope of an empirical science is affected by the mathematics applied to it.

On the other hand, one also searches for general theorems that can be applied to differential equations, or to sets of such equations, when they can't be solved, and which can answer important physical questions despite this. We can wonder, for example, under what situations solutions exist (for what sets of parameters) when they are unique, and how their properties depend on small changes in the values of these parameters. We may also be curious whether there are periodic solutions and, if so, whether they remain periodic when parameters change slightly.[9] Questions about how solutions change as a function of parameters are of great significance because these empirically measured parameters can be only imprecisely specified.

The search for required general theorems takes us quite far. It leads not only to vast developments in highly sophisticated mathematics but to intricate applications of other branches of mathematics (such as topology and abstract algebra) to the subject.

Applied classical analysis is not the only source of tractability problems for empirical science. Any branch of mathematics applied to a subject area brings with it epistemic constraints on what we can know in that subject area because of the unavailability of simple algorithms for deciding general (mathematical) questions. When this happens, a rich and complicated applied mathematics results because we must extract information *piecemeal*, by fashioning special and limited mathematical tools for particular questions. The mathematics applied to contemporary physics provides many examples, from abstract algebra to functional analysis.[10]

I've described briefly one broad collection of methods for extending the scope of a science: the development of sophisticated mathematics which can be used to extract empirical information not otherwise available. Notice, though, not the success of this practice, but rather, that its presence indicates a significant gap between the scope of a science and its program. Epistemically, we can be stopped in our understanding of an area by sheer mathematics, and solutions to the problems posed by this mathematics (when available!) can be piecemeal and arduous.[11]

9 See, for example, Arnold (1980). Computer technology has recently yielded long-term calculations which give fairly robust answers to questions about the evolution of certain *n*-body systems. See Sussman and Wisdom (1992) for results on our planetary system. The task was not easy; but it does indicate that mathematical intractability can sometimes be circumvented by the sophisticated use of computers. Computers are coming to play an equally significant role in the derivation of theorems in pure mathematics also (as philosophers have known for some time). Nevertheless, one should not be sanguine about how far these tools take us. They won't eliminate, once and for all, the problems that mathematical intractability poses for empirical science.

10 See, e.g. Kaku (1993). Also, in the Introduction to Reed and Simon (1978: vii), a work on functional analysis methods in mathematical physics, it is observed that with the publication of the fourth volume, material intended for the first two volumes of a projected three-volume work has been covered.

11 Speaking of the three-body problem, Euler, in his *Recherches sur les irrégularités du*

The use of calculational short-cuts

Any empirical subject matter has limitations in exactness due to the measurement techniques available at the time; any practical application also has thresholds below which errors are insignificant. Therefore, computational short-cuts, which are either mathematically incorrect or empirically inaccurate, are acceptable when they lead only to errors which fall below these thresholds.

Reasoning exemplifying this strategy is commonplace in physics. One finds regularly, for example, the substitution of one mathematical quantity for another because they are close enough in value in the physical circumstances under question to justify the substitution, e.g. the substitution of θ for sin θ when θ is "small," or, more generally, the ignoring of higher-order terms in a Taylor series expansion.[12]

Closely related to this, epistemically speaking, is the approximation of a complicated set of empirical values by a crude numerical constant, for example, the empirically established constants of proportionality in the theory of frictional forces when one dry unlubricated surface slides over another.[13] Another example is the empirically established constant k (for springs of a particular shape, material, etc.) in Hooke's law.

There is no mechanical way of determining when these substitutions or approximations are acceptable. One must have good "physical intuitions,"

mouvement de Jupiter et Saturne, writes: "And if the calculations that one claims to have drawn from this theory are not found to be in good agreement with the observations, one will always be justified in doubting the correctness of the calculations, rather than the truth of the theory" (cited in Wilson 1980: 144.) In his *General Principles of the Motion of Fluids*, published in 1755, Euler says, even more outspokenly: "And if it is not permitted to us to penetrate to a complete knowledge concerning the motion of fluids, it is not to mechanics, or to the insufficiency of the known principles of motion, that we must attribute the cause. It is analysis itself which abandons us here" (cited in Kline 1972: 541). George Smith has pointed out to me that, ironically, the problem in the second case was with the principles of fluids, *not* the analysis itself, although this could not have been realized by Euler at the time.

12 An introduction to this reasoning at its best is Feynman *et al.* (1963). See 7-4, 13-8 (where we are told that $2 \pi \rho \, d\rho$ is the area of a ring of radius ρ and width $d\rho$, if $d\rho \ll \rho$), 27-1, or 29-7 (the first number gives the chapter, the second the page). Feynman won't always point out that he is doing this sort of thing, which can confuse any reader unaware of the ubiquitousness of the approach. These short-cuts are made psychologically natural, in part, by the standard use among physicists of infinitesimals.

13 This theory has two laws: the first holds of static friction (the frictional force operating in resistance to the initial movement of one such object over another), the second of kinetic friction (the frictional force operating in resistance to the continued movement of one such object over another). We have: (1) such forces are approximately independent of the area of contact, over wide limits; and (2) such forces are proportional to the forces exerted by the two objects on each other (for example, if a table is supporting a block of wood).

The constants of proportionality are established by testing the objects in question and determining what (more or less) the frictional forces must be. The micro-phenomena

the sense that the circumstances one is studying will not punish sloppiness, that things are "small enough" or "close enough" to work. Or one can use hindsight, the empirical knowledge that the result yields the "right" answers. Unfortunately, despite its obvious risks, there is often no other way to go.[14]

A technique related to these is the derivation of "laws" which hold provided certain magnitudes (thicknesses, velocities, shapes) do not get "too" small, fast, irregular, or whatever. The technique is familiar from the common use of Newtonian mechanics in contexts (i.e. almost everywhere) where velocities are small enough for errors to fall below significance. But this ubiquitous approach is used in every branch of physics.[15]

The deployment of idealized models[16]

When philosophers speak of idealized models, they often mean frictionless

---

due to one object sliding over another are extremely complex, and so the constants of proportionality cannot be derived from the micro-description of the surfaces, nor the force laws be derived from underlying physical theory. See any first-year physics text, e.g. Feynman et al. (op. cit.: 12-2) or Resnick et al. (1992: 104–8).

14 Hempel (1988: 158) writes of "provisos," which he takes as utilized in the jump from general scientific theory to a specific situation: "The proviso [in the use of Newtonian theory to deduce the state of a binary star system from a previous state] must ... imply the absence ... of electric, magnetic, and frictional forces, of radiation pressure and of any telekinetic, angelic, or diabolic influences." On my view, what is assumed is that such effects – if they exist – fall below desired thresholds, not that they are absent altogether.

15 Here's a neat elementary example. Consider two concentric circular loops of wire, the outer with a slowly changing current which generates a magnetic field, and the inner in which a current is induced. A formula for how the electromagnetic force in the inner loop depends on the changing current in the outer loop is easily derived, provided one assumes the current is changing "slowly enough" and that the inner radius of the loop is "small" compared to the radius of the outer loop. How slowly and how small? This is something one has to learn to gauge in practice: given one's needs, small and slowly enough that deviations from what the formula predicts fall below empirically established thresholds. For details on the derivation, see Purcell (1985: 276– 9). Other examples, such as Ohm's law and self-inductance, are in Purcell (1985: 128–33, 283). Also see Feynman *et al.* (1963: 15-10) for approximating formulas in relativistic dynamics, oscillations in various contexts (23-4, 25-5, 30-11), polarization of light (33-9), and so on.

16 The categories offered here are not disjoint: Some examples of calculational short-cuts can also be seen as using idealized models. Other overlap examples may be found in electrostatics, where standard (illustrative) applications of Coulomb's law or Gauss' law are to things like infinite lines of charge or infinite sheets of charge; see Purcell (1985: Chapter 1), or Resnick *et al.* (1992). Classification turns on degree. I call something a calculational short-cut either if it arises more or less asystematically in the derivation of a result, or if it does not suppress significant phenomena. The physicist sees that he can get from *A* to *B* easily by doing *this*, where *this* is more or less acceptable if things need not be that precise. I describe something as an "idealization" when it turns on systematically excluding phenomena (such as friction) from explicit consideration to make derivations tractable.

planes. But frictionless planes stand at one end of a continuum of techniques that are quite ubiquitous in the sciences.[17]

Consider projectiles (e.g. cannonballs). To calculate the path of one, we can neglect air resistence, irregular distribution of mass in the object, shifts in gravitational force, vibration, rotation, etc. This makes the solution of the problem easily calculable, both in the sense of tractable mathematics, and in the sense that the physical parameters needed are not too difficult to measure. If results deviate too far from the actual path of the cannonball, we can increase the complexity of the physical model at some expense to its tractability (by explicitly acounting for effects of previously neglected physical factors) until – hopefully – we get a model in which the physical parameters are still not too hard to measure, and from which, mathematically speaking, we can still extract information easily; but which gives results close enough to what actually happens to satisfy us.

An important tool for constructing idealized models is the treating of discontinuous phenomena as continuous, so that continuous functions, and the application of integral and differential techniques to such functions, can be used. Obvious examples arise everywhere in physics: in mechanics; electrodynamics (where, e.g., charge is treated as continuous); and in fluid dynamics. Continuous phenomena may also be treated as discontinuous to allow mathematical applications.

Developing sophisticated mathematics to extend the range of cases that physical law can be applied to, introducing calculational short-cuts, and developing simplifying models to bury mathematically and technically intractable aspects of something below desired thresholds, are methods that work together. Newton's theory remained the "hard core" of a successful research program for so long because these methods of extending the scope of Newtonian physics were so successful.

But my concern, I repeat, is not with how far the scope of physics has been extended since Newton's time but rather with how, even today, its scope is vastly narrower than its program. All the different sorts of idealization, and their accompanying heuristics, have one purpose: to extend the scope of scientific law without having to account *explicitly* for all the pertinent aspects of the thing studied.[18] Sometimes the idealizations in question are not inevitable, practically speaking: one uses them because they make the application of scientific law a little easier, and one can dispense with the short-cuts if challenged. But an equally common case is one where

---

17 Suppe (1989b: 65–6) alludes to examples indicating use of idealizations in physics, chemistry, biology, psychology, linguistics, and anthropology. Also see Achinstein (1987, 1990) for an extended discussion of Maxwell's use of idealizations in his derivation of the gas laws from Newtonian kinematics.

18 Feynman *et al.* (1963: 2-2) puts it this way: "[W]e can often understand nature, more or less, without being able to see what *every little piece* is doing ..."

idealizations are present because no one sees how to apply the laws directly to what they are supposed to apply to: The heuristics are essential.[19]

In general, when introducing computational tricks or idealizing models to bridge the program–scope gap, there is often more than one way to do it. Worse, alternative methods need not give the same answers.[20] In the elementary texts I've been primarily citing, one sees approximation methods in contexts where they are (pretty much) successful – and where, since the mathematics being used and the phenomena being idealized away are fairly well understood, physicists usually have a good sense of what is being left out. In journals and in advanced texts, approximation methods of the same sort are used in far riskier circumstances both as regards the mathematics and as regards the phenomena in question.[21] But their philosophical significance, regardless, is the same.[22]

The design of laboratory experiments

The fourth indicator of the program–scope gap in the sciences is the ubiquitous need to test scientific claims in experimental situations. The point of an experiment is similar to that of idealization. But idealizations are applied in situations where the complexity of the subject matter is ineliminable. Experiments, by contrast, are commonly the *creation* of situations wherein the undesirable complexity has been either minimized (below a certain significant threshold) or eliminated altogether.[23]

---

19 For example: in Sussman and Wisdom, we find (1992: 57):

> Our physical model is the same as that of QTD except in our treatment of the effects of general relativity. General relativistic corrections can be written in Hamiltonian form, but we have not been able to integrate them analytically. Instead we used the potential approximation of Nobili and Roxburgh ... which is easily integrated, but only approximates the relativistic corrections to the secular evolution of the shape and orientation of the orbit.

Also see Achinstein (1987: 242–3).

20 See the discussion in Cartwright (1983: 78–81) of six approaches (involving three distinct Schrödinger equations) to the theory of quantum damping, and its associated emission line broadening phenomena.

21 A nice book for this is Kaku 1993: He often points out where the risks are, both historically, in the development of quantum field theory, and presently. Original papers are cited fairly regularly as well.

22 Cartwright is sensitive to the widespread presence of these methods in physics. See, for example, her detailed discussion of how approximation methods are used to derive the exponential decay law in quantum mechanics (1983: 113–18), her illustrations of how approximation methods can yield different answers, and how they are empirically adjusted (for idealized models applied to amplifiers, pp. 107–12, and for mathematical short-cuts used to analyze the Lamb shift, pp. 119–23), etc.

23 Here are three things an experiment can get us. First, where more than one factor is at

For example, in dropping items from leaning towers, one may neglect the impact of air-resistance or approximate it by a crude numerical constant. But when testing the rate at which objects fall in an artificially created vacuum, the undesirable complexity introduced by air-resistance is largely absent, and consequently one can apply (certain) physical laws directly to the situation. A scientist can test scientific claims in experiments wherein the test situation is artificially simplified in this way; and, naturally, enormous sophistication and expense can go into the design and implementation of experiments. The result, when the experiment succeeds, is the creation of an unusual locale where the parameters allow relatively direct application of science to the phenomenon: We can predict an outcome and be reasonably sure that, if it fails, the fault is in the body of science so applied.[24]

But again, notice how the need to create such experimental environments usually indicates a program–scope gap: The intractability of most ordinary situations prevents any direct test of scientific claims by applying them to, or testing them in, such situations.

The use of formal heuristics

I separate this last group of approaches explicitly from previous ones despite the fact that, in practice, there are examples of formal heuristics rather close to examples from the earlier categories. The reason for the separation is that physicists have a different view of cases that clearly belong here than they do of the others.

I start with an historical example, or with how this particular historical episode is often described. Ptolemy and his successors are often taken to

---

work, we may be able to isolate the factor(s) we are interested in. Second, even where only one factor is involved, we may be able to reduce "background noise" so that the effects of the factor can be more easily recognized. Finally, and this point is related to the last one, an experimental situation may be designed that can cater to the measurement techniques available at the time. If we want to measure the acceleration on objects due to gravity, it may be beyond us to measure it by means of falling bodies. But if objects can be made to accelerate more slowly (e.g. by placing them on an inclined plane), significant and stable measurements of acceleration may become possible.

And notice: although the use of experiments often indicates a program–scope gap, this isn't *always* the case. Sometimes an experimental situation is designed simply to *explore* how something operates when certain factors are missing (e.g. gravity-free environments).

24 Such direct application of science to a situation does not mean there are no additional assumptions involved in the experiment which cannot be derived from the scientific theory applied, or from science as a whole. Indeed, there often are assumptions that materials or the apparatus used have certain properties. These assumptions are often based on gross regularities, not scientific doctrine, as I show later.

By no means am I suggesting that it is easy to tell what the results of an experiment are due to. Matters can be quite delicate. See Franklin (1993) for an extended case study of this.

have been concerned with "saving the appearances":[25] that is, their astronomical constructs designed to predict the movement of the planets were not taken to describe physically real movements of those planets. Rather, their purpose was to supply an easily used mathematical instrument to predict results within the observational standards of the time. The requirement on contemporary uses of formal heuristics is the same: to find a mathematical model which predicts certain aspects of the subject matter being modeled to within specified significant figures, without requiring that details of the model are realized in the subject matter; hence my stress on the "formal" in "formal heuristics."

Physicists *epistemically* distinguish this approach from the four earlier approaches. The point of idealization, when it is *properly* idealization, is that one knows, at least roughly, what is being idealized away from. When frictional forces between surfaces are neglected or approximated by a numerical constant, the physicist takes herself to know, at least in a general way, what is being left out, and why it is being left out. Furthermore, in many cases, she knows that the development of more powerful information-gathering techniques, or mathematics, may enable the inclusion of aspects of the subject hitherto neglected. A similar claim holds for computational short-cuts. Where the mathematics is fairly well understood, one knows roughly *what* is being idealized (what geometric irregularities are being ignored, etc.), and one knows what one is facing should these complications be included. One also knows the sources of the complications.

But with formal heuristics, an instrumental mechanism that gives answers of a certain sort is introduced out of whole cloth. Although it yields the right answers (to within certain parameters), we may know that it is doing this via a physical description that has nothing to do with what's really going on physically;[26] or we may have no idea at all of the physical mechanisms actually at work. In the Ptolemaic case, when significant deviations from empirical measurements arose, one had no idea where they were coming from (one couldn't even guess whether the deviations had a mathematical source or a physical one). One, therefore, didn't have a clue where to look for reasonable corrections.[27]

---

25 Cartwright (1983: 90) cites Duhem (1969: 82) where the latter cites Piccolomini:

> for these astronomers it was amply sufficient that their constructs save the appearances, that they allow for the reckoning of the movements of the heavenly bodies, their arrangements, and their place. Whether or not things really are as they envisage them – that question they leave to the philosophers of nature.

26 Kepler's "vicarious theory" is an example of an instrumental method for getting data (the heliocentric longitudes of Mars) which he knew to be false in its description of the movement of the planets, but that he could use to establish a theory he could believe. See Wilson (1972).

27 Despite the appearance of being an historical reference, what I say here, and what Piccolomini

Unsurprisingly, formal heuristics usually arise at the cutting-edge of research, where, typically, the subject matter is so complicated or badly understood that scientists have little idea what is going on physically or mathematically. The application of formal mathematics which yields empirically acceptable predictions enables one to carry on (and to gather data) until something physically and mathematically more substantial comes along.

For this reason illustrations of formal heuristics in elementary texts are rare. But here are two examples from Purcell (1985). First, he analyzes magnetic fields in matter by treating matter as if it contains "small" bands of current (even where it clearly does not). The approach is motivated nicely:

> Consider first a slab of material of thickness dz, sliced out perpendicular to the direction of magnetization.... The slab can be divided into little tiles. One such tile, which has a top surface of area da, contains a total dipole moment amounting to $M\,da\,dz$, since M is the dipole moment per unit volume.... The magnetic field this tile produces at all *distant* points – distant compared to the size of the tile – is just that of any dipole with the same magnetic moment. We could construct a dipole of that strength by bending a conducting ribbon of width dz into the shape of the tile, and sending around this loop a current $I = Mc\,dz$.... That will give the loop a dipole moment:
>
> $$m = \frac{I}{c} \times area = \frac{Mc\,dz}{c}\,da = M\,da\,dz$$
>
> which is the same as that of the tile.

(1985: 423)

Purcell later sketches the trick of modeling a uniformly magnetized rod by

said, seems unfair to Ptolemaic astronomy as practiced. When justifications are offered by, e.g. Ptolemy, they look like this: "Now it is our purpose to demonstrate for the five planets, just as we did for the sun and moon, that all their apparent anomalies can be represented by uniform circular motions, since these are proper to the nature of divine beings, while disorder and non-uniformity are alien [to such beings]" (cited in Hatfield 1990: 153, note 22; brackets his). Although Hatfield argues that it may be a bit much to attribute substantial metaphysical beliefs to Ptolemaic astronomers on the basis of such statements, it does seem they were operating with a set of specific mathematical assumptions, e.g. that the motions appearing in the sky are composed of circular motions, that the mathematics they employed is exemplified in celestial motion, etc., assumptions which made it clear to them exactly where the deviations were coming from, and what, broadly speaking, should be done to institute corrections. They were not "saving appearances." Hatfield (1990: 155 note 37) writes: "Much mischief has been caused by the acceptance of Duhem's characterization of Ptolemaic astronomy as inherently 'instrumental' ...".

means of "fictitious" positive and negative magnetic charges (magnetic monopoles). He writes: "You may want to use this device if you ever have to design magnets or calculate magnetic fields." But the next paragraph begins: "We must abandon the magnetic pole fiction, however, if we want to understand the field inside the magnetic material."[28] The subsequent discussion that is supposed to yield understanding of the magnetic field inside the material is idealized, too: the field itself is an idealization much the way frictional forces are (a point stressed by describing the field as strictly macroscopic), and the technique of modeling the mathematical properties of the field by treating it as containing "small" bands of current is used (since the magnetization yields the same field as a band of current, it is clear that the heuristic of bands of current is used to understand what sort of *field* results inside the magnetic material rather than to describe from what it is arising).

The beginning physics student is guided to understand the physical status of different idealizations, what they idealize away from, when that is how they are operating, and how they yield what we need, if they are entirely fictitious: This is done against an implicit mathematical and physical background which is more or less well understood (at least by professionals).

For illustrations where things are not well understood, see Cartwright (1983: 129) on treating radiating molecules in an ammonia maser as if they are classical electron oscillators, or her discussion of the application of the harmonic oscillator model in quantum mechanics, "even when it is difficult to figure out exactly what is supposed to be oscillating ...".[29]

I've illustrated these five methods of applying and confirming scientific theories with examples from physics. My choice of illustration shouldn't leave the reader with the mistaken impression that the program–scope gap is restricted to physics, and branches thereof, since it arises in the special sciences, too.

The program–scope gap: illustrations from the special sciences

Here's an example from chemistry. Chemical stoichiometry is the study of

---

28 Purcell (1985: 430). Purcell's language is quite misleading. I don't think he means to suggest the magnetic-monopole fiction enables us to *understand* the magnetic field *outside* the magnetic material, if "understand" is to have the implication that we are to believe the magnetic field *really is* due to magnetic monopoles. Monopoles are being offered (to budding engineers) only as a calculational device.

29 Cartwright (1983: 145). She regards the repeated application of this model in cases where there is no real underlying physical understanding of why the model yields the right empirical results as one that "gives explanatory power." This is wrong: in most of these cases an explanation of why the harmonic oscillator model can be applied is precisely what we *don't* have. I discuss the matter further in § 5; but let me say now that explanation in the empirical sciences is pretty clearly linked (by physicists) to realist presuppositions.

quantitative relationships among reactants and products in a chemical reaction. The theory is solid and routine: the well-established chemical formulas for certain substances (and atomic-mass figures for the elements) allow translations to and fro of mass measures of substances into molar measures (numbers of molecules) – and these enable predictions of the quantitative outcomes of various reactions.[30]

However, there is a distinction between "theoretical" and "actual" yields – for purity of substance is an ideal almost always out of reach, reactions are sensitive to temperature, pressure, and other variables, there are always unexpected variations due to the methods of executing the reactions (e.g. losses due to the transferring of substances from one container to another), not to mention out-of-the-blue surprises. As any elementary chemistry textbook points out (so novices don't suffer from unreasonable expectations), actual yield is an experimentally determined number that *cannot* be calculated beforehand, and one sensitive to changes in the methods of carrying out the reaction process.[31]

One might argue that this example doesn't illustrate the program–scope gap. For perhaps the theory behind chemical stoichiometry is confirmed not in the direct way one might have naively expected but in a more sophisticated way that turns on the simultaneous application of other subsidiary theories about materials used in the reaction process, temperature changes, and so on. But then, the objection goes, what we have here is a special science whose laws apply to the phenomena they apply to *in conjunction with* laws from other sciences.

Although this is fairly common – the laws of several sciences are applied to the same domain – I don't think this is going on here. While actual yield isn't derivable from the theory in question, it isn't derivable from the theory in question plus "subsidiary theories" *either*. In calling the result "experimentally determined," the point is that actual yield is not derivable from a collection of theories *at all*. No one should mistake an *ex post facto* explanation of why empirical yield deviates from theoretically predicted yield for the application of further theory. It is missing the crucial element needed in the application of theory: our capacity to predict deviations before actually being faced with them.

Here's another of the many examples of this sort of thing in chemistry. A common practice is to read off the properties of chemical substances from the geometric pattern of their constituents in space; and the method involves great idealization. The properties such models attribute to molecules are not derived from the underlying quantum-mechanical laws but rather from relatively simple repulsion and attraction models, utilizing simplified three-dimensional shapes. Higher-level forces emerge, such as torsional energy, van der Waal's forces, dipole–dipole interactions, hydrogen bonding, and so on.

---

30 I have simplified my exposition slightly to avoid a discussion of ions that changes nothing essential to the point.
31 For example, Stoker (1990: 328).

These mathematized models are greatly limited in scope relative to their programmed range of application in *chemistry*, and *the extent of these limitations is determined* (for the most part) *empirically*. That is, when faced with a new reaction, such models are *at best* rules of thumb for predicting the mechanisms of the reaction (and therefore its properties vis-à-vis other solvents, temperature changes, and other variables).

The program-scope gap: concluding remarks

The program–scope gap helps explain why empirical science is so hard to do. The impression that physics, chemistry, and other ("hard") sciences have research directions relatively immune to sociological factors, such as institutional bias (in contrast to "softer" sciences), is best explained by the fact that extending the scope of such a science is arduous.

Here's an illustration. Schrödinger's equation describes the evolution of a system when the system's kinetic and potential energies have been represented in a mathematical object called the *Hamiltonian* of the system. Cartwright (1983: 136) points out that students of quantum mechanics learn to manipulate Hamiltonians for a small number of highly idealized situations, and then learn to extend the scope of such Hamiltonians to other situations by combining the original ones. But why are there so few to begin with? Cartwright offers a Kuhnian explanation:

> The phenomena to be described are endlessly complex. In order to pursue any collective research, a group must be able to delimit the kinds of models that are even contenders. If there were endlessly many possible ways for a particular research community to hook up phenomena with intellectual constructions, model building would be entirely chaotic, and there would be no consensus of shared problems on which to work.
>
> (1983: 143–4)

And a little later, she adds:

> A good theory aims to cover a wide variety of phenomena with as few principles as possible…. It is a poor theory that requires a new Hamiltonian for each new physical circumstance. The great explanatory power of quantum mechanics comes from its ability to deploy a small number of well-understood Hamiltonians to cover a broad range of cases, and not from its ability to match each situation one-to-one with a new mathematical representation. That way of proceeding would be crazy.
>
> (ibid.: 144–5)

But this gives an entirely false impression of the motives here. The way of proceeding Cartwright describes is required *only because* the mathematics of Schrödinger's equation and the Hamiltonians is so hard. If it were *easy* to construct the Hamiltonian for any situation, and easy to derive its evolution via Schrödinger's equation, then the "crazy" method would be just the right one.

Two last points. The reader may wonder if the scope limitations previously illustrated imply that the program of a science, as I understand it, is unduly reified. Perhaps practical limitations should be taken to indicate that the program of a science, when faced with such, is to be narrowed to within the parameters of the scope of our tools. This is an unreasonable suggestion because the scope of any given science is constantly changing, and in a way that cannot be systematically described. Mathematical advances, instrumental developments, increases in computational speed, etc., can shift scope drastically without affecting the structure of the scientific theories so applied at all. The "programmatic talk" of scientific theories must always, for epistemic reasons, outstrip what our actual evidential practices scope out (as manifested in applications and tests). We will see, again and again in this book, this contrast between language, and the technical epistemology that buoys it.

Second point. Some may feel I've illegimately taken a realist view towards scientific law in my discussion of idealization. An alternative view takes scientific laws to apply perfectly to idealized objects, rather than taking laws to apply to actual objects, but in a way obscured by "noise." I'm not sympathetic to an instrumentalist view of scientific laws – at least when it comes to fundamental laws – for two reasons. The first is that, as I've indicated, ordinary scientific practice reveals different attitudes to idealizations that are used purely instrumentally as distinct from those cases where they're intended to surmount forms of intractability. Second, I think a realist view – of a sort – can be philosophically defended, as I attempt to do in § 5.

§3

REDUCTIONISM,
CONFIRMATION HOLISM,
THEORETICAL DEDUCTIVISM

I've argued for the program–scope gap in science by showing how common scientific methods presuppose it. I now describe its presence differently, by bringing it to bear on certain philosophical doctrines. I first show how the irreducibility of most of the laws and notions of the special sciences is due to the program–scope gap. Then I describe how the gap reveals limitations in confirmation holism. Finally, and most importantly, I indicate how the gap affects how scientific theories are tested and applied.

One caveat. The reductionist literature is vast. My aim is, however, narrow: to illustrate how the program–scope gap illuminates issues here; and so I touch on reduction, for the most part, only in so far as it bears on this.

Consider, then, the project of reducing ("defining in terms of" and "deducing") the terminology and laws of one or another special science to physics. It is in each case the *scope* (not the program) of physics that decides the fate of this project. For to define the vocabulary of a special science in physical terms, one must apply physical law in enough detail to the domain in question to recognize the differences the special terminology marks out and to derive the special laws, if any, that hold in that domain. This makes *tractability* a requirement on any attempt to reduce the laws of a special science to one that (metaphysically speaking) lies under it.[1]

Here's an illustration of how fast tractability problems arise, and how strongly they flavor a special science. Even a shallow perusal of the philosophical literature on the mind–body problem[2] gives the impression that some philosophers think a crucial block to reducing psychology to biology

---

1   The physical distinctions, mentioned in the digression in § 2, that can be used to demarcate the domains of the special sciences, do not by themselves explain the presence of such sciences. Their presence is explained by epistemological scope problems in the underlying sciences; for if such problems did not exist, reductions of the notions and laws of the special sciences would be possible, despite physical distinctions in subject matter; and in such a way that application and testing of the reduced theory would be unaffected.
2   For samples, see articles in Block (1980 and 1981), and Rosenthal (1971 and 1991); see also Fodor (1979). I discuss issues of reduction as portrayed in this literature a bit more in § 4.

(or, more specifically, to neurophysiology) is the non-existence of type–type reductions of psychological predicates to biological ones. So, one *might* think everything is available for successful type–type reductions of the terms and laws of *chemistry* to those of physics. Chemical kinds can be defined explicitly in terms of physically different molecules,[3] and, theoretically, Schrödinger's equation implies the chemical properties of any substance from a description of its physical make up.

Nevertheless, chemical laws remain a bewildering mixture, only *some* of which can be derived from the underlying physics – and a good number of these only in special cases. This is because Schrödinger's equation for atoms containing more than one electron cannot be solved in closed form: Numerical methods providing approximate solutions must be employed if quantum mechanics is to bear on the structure of the atom.[4]

A similar situation obtains for many of the notions in chemistry, as indicated in § 2: they enjoy a semi-autonomous status. They cannot be derived directly from underlying quantum-mechanical notions, but instead their application is supplemented, *where possible*, by underlying physical laws. This semi-autonomous status is indicated also by a kind of open-endedness of the notions. For example, the notion of *bonding*, crucial to chemistry, can suddenly mutate when substances are discovered whose constituents do not share or exchange electrons in the extant list of ways this can be done.[5]

To some extent, I have made points philosophers are familiar with: thus, much of the terminology and the laws of the special sciences cannot be reduced to physics, or to other special sciences that (metaphysically) lie under them; and intractability (from the physical viewpoint, and from the viewpoint of the underlying sciences) is what motivates the autonomous elements in the technical and theoretical practices of a special science.[6] But notice two points that have not been widely observed, and which follow directly from a recognition of the program–scope gap. First, confirmation holism does not impact on scientific practice as much as philosophers of science have assumed. Second, scientific theories are partially insulated from what

---

3   This, in fact, is the job of the periodic table, although, of course, it doesn't tell the whole story.

4   Recall the discussion of differential equations under *development and application of sophisticated mathematics* in § 2.

5   Open-ended notions in special sciences remote from physics are commonplace (think of the notion of *government* in linguistics, or, in biology, such notions as *virus* or *fitness*). One mark of this open-endedness is that definitions of these notions are not available – only paradigmatic examples. It should be no surprise the notions of a special science act this way; the surprise is that this open-endedness, and failure to be definable in physical terms, is present in some of the notions of *chemistry* where the project of a type–type reduction (of the sort sought for psychological predicates by physicalists, for example) has been successfully carried out for chemical kinds.

6   See Fodor (1979) and Putnam (1975b), for early examples.

would otherwise be their appropriate domain of application; there is, as I like to put it, a *boundary layer* between a scientific theory and the domain it is taken to apply to. I develop these points in turn.

The program–scope gap mutes confirmation holism by isolating branches of the sciences from one another: one can be impressed with how much of physics can be brought to bear on chemistry; but equally impressive is how much chemistry must be practiced without the benefit of physics because physical law cannot directly apply to chemical situations.

Another example: one may, as Chomsky seems to, consider linguistic competence to be a biological (i.e. neurophysiological) phenomenon. That is, one may take the program of neurophysiology to include phenomena now studied by syntacticians. So much, of course, is theory – or barely that, *programmatic* hope. How much confirmation holism lurks here, *in practice*, depends not on the program of neurophysiology but rather on its scope. And, consequently, there hasn't been much *at all*. Few, if any linguists, can rely on anything more than speaker–hearer intuitions for (real) data.

Turning to the second point, the physicalist program decrees that physical laws, in principle, apply everywhere. But it is the scope of physics that actually dictates where *evidence* can have its impact on physical theory. For despite all the work generating idealized models and sophisticated mathematics, it is only sporadically, and under special circumstances, that physically tractable situations arise (even *modulo* idealization), and it is only in physically tractable situations that we can *test* and *apply* physical laws. For recall one thing that is required: the impact of the mathematically and physically intractable aspects of a phenomenon must fall below a certain threshold. Alas, most of what we see in the world does not fit this category (e.g. *most* of the movements of, and in, animate beings), and quite a bit does not fit it particularly well (e.g. air-flow). This is why applying and testing physics is so *hard*. The piecemeal fashioning of limited solutions based on limited information to fit special and complex situations *just about* guarantees that any progress will be rather slow, and marked by the kind of brilliant insight that (when required of scientific practitioners) indicates we are in an epistemic jam.[7]

Why has this obvious implication of the program–scope gap been nearly universally overlooked by confirmation holists? The holist is impressed, much like Quine, by the fact that statements do not meet the tribunal of experience alone. He explains why theories do not directly confront their

---

7   Kuhn (1970) shows sensitivity to this effect of the program–scope gap because it is tractability problems that both isolate a science from what would otherwise be its appropriate stomping grounds and explain the existence of the problem-solving practices in what he calls normal science. Nevertheless, in my view, Kuhn does not give the program–scope gap its proper weight.

intended domain of application solely by the fact that other theories are brought to bear on that domain as well.[8] Theory A does not directly bear on its intended domain because it is applied in tandem with theory B; and we could falsify aspects of B instead of A. But in explaining *all* cases this way, he has failed to recognize cases where a theory's inability to directly confront its intended domain is due not to its interfacing with other theories but simply to a straightforward program–scope gap.

This is not all that's missed. Another common tendency is to adopt an oversimplified view of how scientific theories get tested or applied, what I've called *theoretical deductivism*:[9] the view that all that is involved in testing or applying a scientific theory is directly deducing consequences from it. Even with the qualification that such deductions take place, generally, not within the confines of one theory but in conjunction with other theories (as the contemporary confirmation holist will stress), the view is still largely wrong.

Why? A quick answer is available from § 2: a program–scope gap prevents us from applying scientific doctrine in the kind of detail needed to give us a step-by-step algorithm for designing and implementing an experimental test or product. Recall the importance of idealizations and computational short-cuts. The uses of such devices occur not on the pages of scientific dogma, but in the margins. For it is often not part of scientific theory, strictly speaking, to indicate how falsifications can be introduced via calculations and idealizations to yield errors falling below desired thresholds.[10]

The deductive holist may be undisturbed by these considerations; he might claim that holism requires the *entire* web of belief to be the pertinent applicable theory, and so he might think the needed consequences *are* deducible from that (entire) web – which includes the marginalia just referred to. This response misses the point, because the reason it is not part of scientific theory, strictly speaking, to indicate how falsifications can be introduced in calculations and idealizations to yield errors falling below desired thesholds is that there is little that can be said, in general, about how to employ such cook-book tricks. Approximations, for example, can be introduced in surprising and unexpected ways that one can discover only by sheer guesswork. To suggest that "employing an approximation to a formula will make the mathematics more tractable while still yielding results below

---

8 Recall this interpretation of chemical stoichiometry at the end of § 2.
9 Among the theoretical deductivists are Churchland, Duhem (1914), Feyerabend, Lakatos, Popper, Putnam, and Quine.
10 "Not usually," doesn't mean "never." Sometimes one *can* calculate exactly how much error should result from using an idealization, and the fact that the error resulting from the idealization matches the calculation is itself further confirmation of the theory, or the theory in conjunction with some further background theories. But, often, as we have seen, that the idealizations work is something that has to be established empirically rather than deduced from theory.

measurement thresholds" is part of our web of belief that enables us to "deduce" certain results so evicerates the notion of deduction that the deductive holist intent on this move may be said to have emptied his claim of any genuine methodological content.[11]

Anyway, even in conjunction with successfully applied heuristic devices, such as those considered in § 2, theory – however broadly understood – stops well-short of being a sufficient guide or recipe in the experimental situation. This is because experimental situations in which scientific theories are generally tested, or new products developed, are not ones whose full descriptions are derivable from either the scientific theories directly involved or such theories in conjunction with auxiliary theory plus heuristics; for all this is usually still seriously incomplete. Here are some illustrations of how theory + heuristics can fall short as a guide in the experimental or product-development situation.[12]

First, we may be trying to create an experimental situation where certain complexities are minimized. The theory, even with good company, may have nothing of interest to tell us about how far the minimization process can go before we get results we like; it may have nothing of interest to tell us about how the experimental situation may have gone wrong, if it goes wrong, or that it is possible, given the apparatus in question, to get what is experimentally needed. In such a case, we have to tinker with the apparatus until we get it right (or give up), and we may never have much of an idea, one way or the other, for why it went wrong the way it did.

For example, we may test a theoretical claim that cathode rays are deflected by an electric field in a vacuum. But this claim, even with company, may tell us nothing about how close to a pure vacuum the experimental apparatus must achieve before we see the predicted results; it may tell us nothing about how to tinker with the apparatus to get what's needed, or even how, after tinkering, to recognize how much of a vacuum we've got.

Second, the experimental situation can be one which clearly tests the theory in question or one to which a theory clearly applies, but not one the description of which, even in broad terms, is derivable either from the theory in question or from that theory in company with subsidiaries. Usually, the situation is too specific to be so derivable: One regularly relies on all sorts of specific know-how in the design of experimental apparatus or products, and one need not have much by way of theory that explains how the stuff used works.[13] One example is the glass lenses in the telescopes used by

---

11 I am leaving aside, for the moment, a second objection to the deductive holist – that he so broadens the notion of "theory" by packing everything in our web of belief into it that he overlooks the crucial epistemic difference between generalizations arising from theoretical deductions – in the appropriately narrow sense of "deduction" – and generalizations arising empirically – what I call "articulated gross regularities." I discuss this difference in §6.

12 I am indebted to Achinstein (1991: 308–17) for some of what follows.

13 Of course, we can subsequently learn a lot about the materials used to construct scientific

seventeenth- and eighteenth-century astronomers in the absence of *any* theory of the structure of glass that could be relied on to explain how glass could do the job needed. Another detailed example is afforded by Collins (1992: 51–78): he describes the attempts of the trained scientist Bob Harrison to build a TEA-laser of the double-discharge design. Crucial to its design and eventual successful implementation was a great deal of know-how about how electrodes should be connected, how long they should be, what components of the laser should be insulated, and how, and so on, all of which must be empirically established in the very context of building and testing the laser itself.

The point, then, is this. As I have claimed, there is often a boundary layer between a theory or group of theories and the subject-matter we attempt to apply it (or them) to, and I've provisionally indicated its nature by illustrating briefly how much empirical work is still needed to implement an experiment or product, even when the scope of a theory is augmented by conjoining it with additional theories or by applying to it the various heuristic devices described in § 2. This boundary layer is overlooked by theoretical deductivists, in part, as I have argued, because they're willing to claim that any generalization used is part of the theory so applied, *even if the generalization is one that must be derived not from the web of belief being tested, but in the process of applying that web of belief to the very situation that it is being applied to.*[14]

I don't disagree with confirmation holists who stress the programmatic aspects of the sciences in contrast to their scope, for I don't urge a return to the good old days when statements honestly wore their confirming instances on their sleeves and bravely stood alone against the onslaught of experience. Nor do I think there are *principled* boundaries marking out the data appropriate for any scientific theory, since we cannot stipulate ahead of time what the scope of any scientific theory will be, or in what unexpected direction it might suddenly extend. We can't, that is, even be sure that what looks like ineliminable limitations in scope actually are such.[15]

One might say that confirmation holists have never claimed more. Confirmation holism is a thesis about how conceptual schemes change, and

---

devices. But, almost invariably, we cannot derive all our knowledge of the specific details of how the materials used operate from auxiliary scientific theories about them, or even put such knowledge into an articulated form.

14 In § 7, I discuss how this leads the deductive holist to fail to see any borders between evidential practices that operate independently of the empirical and mathematical sciences, and scientific practices proper. Quine, Lakatos, and Churchland are all prominent examples.

15 Three bodies seems to be a natural dividing-point in celestial mechanics between where general solutions in closed analytical form are available and where numerical approximation methods are required. But *perhaps* changes in our physics or mathematics will make the mathematics for n-body problems trivial. Not likely, of course.

the slogan is "Revision *can* strike *any*where (and from any direction)"; *not* "Revision *does* strike *every*where (and on a regular basis)." But philosophers have underestimated the significance of this modal distinction: scientific evidential practices and, with them, the areas of successful application are, respectively, designed, and marked out, not so much on the basis of the insight that evidence for a theory could come from anywhere, as on the fact that where evidence for a theory does come from is largely determined by the tools, theories and methods currently available.

This means epistemic practices in science are influenced by our best *current* scientific views about which methods work for what subject-matters; and consequently, the program–scope gap has epistemic clout. It is especially the broader considerations of the scope limitations that arise because of intrinsic limitations in our technical and mathematical tools that bear on the breadth of confirmation holism (how large the "wholes" involved are when change in scientific doctrine occurs).

Notice the difference. The confirmation holist sees confirmation forcing large units, such as different theories, to be tested together against experience. By contrast (one of degree), I find tractability problems isolating theories from each other, and creating boundary layers between theories and the domains they are supposed to apply to. Indeed

(a) special sciences develop bodies of methods that are, to some extent, autonomous from the sciences we would like them to be reducible to; and

(b) much of the evidence gathering that goes on in the lab, and much of the application of scientific doctrine to the world, goes on in an empirical way.

One *tinkers* one's way to the point where scientific theory can be applied effectively.

Confirmation holists lose sight of these facts because they explain certain phenomena as due to confirmation holism when in fact they are directly due to the program–scope gap. As a result (§ 6), they also overrate the impact of scientific change on other aspects of our practices and beliefs.

§4

SOME OBSERVATIONS ON REDUCTIONISM AND THE "AUTONOMY" OF THE SPECIAL SCIENCES

My belief that we can more easily explain the presence of special sciences once we deny deductive holism might make it reasonable to predict a sympathy with claims that the special sciences are, in the language of certain philosophers, autonomous;[1] where views about the ways one science might reduce to another, and issues about the supervenience of the terminology of (or the properties studied in) a special science on the terminology (properties) of an underlying science, affect what "autonomy" means.

However, I'm *not* sympathetic with how the relationship between sciences is portrayed in this literature. The difference in view is worth pursuing, first, because it will make clearer what considerations *are* pertinent to evaluating the relationships between sciences, and, second, because the points I make in the current polemical setting are used constructively later.

Although I've concentrated in the foregoing almost entirely on two sciences, physics and chemistry, I nevertheless now argue that issues about reduction are pretty similar with regard to any special science, regardless of whether the concern is with chemistry and physics, or with psychology and neurophysiology.

Much of the literature involves reduction in the following sense. Given two families of terms: $\mathfrak{S}$, and $\mathfrak{R}$, and two sets of laws, $L\,(\mathfrak{S})$ and $L\,(\mathfrak{R})$, couched in the terminology of $\mathfrak{S}$ and $\mathfrak{R}$, respectively, and which imply existential commitments to their respective terminologies, the aim is to provide a translation of sentences in the vocabulary of $\mathfrak{S}$ to the vocabulary of $\mathfrak{R}$, so that translations of the sentences of $L\,(\mathfrak{S})$ follow from the sentences of $L\,(\mathfrak{R})$. I call these "derivational reductions."[2]

---

1 E.g. Dupré (1993), Rosenberg (1985).

2 Reduction thus understood turns on (1) the terms in $\mathfrak{S}$ and $\mathfrak{R}$, and (2) the sets of laws $L\,(\mathfrak{S})$ and $L\,(\mathfrak{R})$, couched in those respective vocabularies, and not, say, on there being different logical connectives involved.

   Definitions connecting the two sets of vocabulary are often called "bridge laws" or "bridge principles," after Nagel (1961), and they may be not actual definitions of one sort of thing in terms of another but instead empirical law-like correlations. For our purposes, it won't matter.

38

Derivational reductions fit best in mathematics, where we find many successful cases of them. Think of the "reduction" of number theory to set theory, for example, where the set of axioms of first-order Peano arithmetic is L ($\mathfrak{I}$) and the set of axioms of ZFC is L ($\mathfrak{R}$). Reductions here are explicit definitions of the vocabulary of $\mathfrak{I}$ in $\mathfrak{R}$ and the resulting translations of sentences in L ($\mathfrak{I}$) follow deductively from sentences in L ($\mathfrak{R}$).

One reason for the compatibility is that axiomatization is rampant in mathematics. Sets of laws (e.g. Newton's laws of motion) may do a similar job if the science in question is highly articulated. But elsewhere (e.g. in biology) one must replace sets of laws by far more amorphous collections of statements in the science that we both take to be true, and to shed light on why other facts in the domain of that science are true. I largely waive this worry in what follows, although it bears significantly on the question of what sense exactly can be made of one or another program of (derivational) reductionism.

Most philosophers, however, require derivational reductions to be subject to additional requirements if they are to work in the empirical domain – I discuss this presently.

The presence in or absence from science of (appropriately restricted) derivational reductions is taken by many to (a) carry ontological weight, and (b) bear on scientific practice. Regarding (a), a derivational reduction is seen as committing us only to what the terminology $\mathfrak{R}$ refers to; the failure of such a reduction is seen as committing us to what both the terminology $\mathfrak{R}$, *and* the terminology $\mathfrak{I}$, refer to.[3]

The bearing of appropriately restricted derivational reductions on scientific practice emerges if we accept the claim that confirmation can work only via *projectable* predicates:[4] without a derivational reduction of projectable predicates to projectable predicates, "property reduction" fails, that is, the properties of an overlying science (which, say, the primitive vocabulary of $\mathfrak{I}$ picks out) do not reduce to the properties of the underlying science (which, say, the primitive vocabulary of $\mathfrak{R}$ picks out); or, to put it another way, the

One can also adopt eliminativism by allowing the vocabulary of $\mathfrak{I}$ to be replaced by alternatives defined directly from $\mathfrak{R}$, and by allowing the set of laws L ($\mathfrak{I}$) to be replaced by a set of laws or truths L' ($\mathfrak{R}$) derivable from L ($\mathfrak{R}$). Mixed positions are also available; see Dupré (1993) for further details on these possibilities. I hereafter describe these positions indifferently as "derivational reductions."

3 When Dupré (1993) claims that various reductionist programs, relying on one or another species of derivational reduction, fail, he has metaphysical ambitions: he wants to show that in some sense, a "robust monism" is wrong.

   The literature is not as explicit as one could wish on exactly what kind of argument is being used here. I conjecture that, as with the Quine–Putnam indispensability thesis in the philosophy of mathematics, implicit use is made of a Quinean criterion of ontological commitment.

4 The classic discussion of projectability and confirmation is found in Goodman (1955).

kinds of the overlying science do not reduce to the kinds of the underlying science.

This additional requirement on derivational reductions has bite in the philosophy of both biology and mind because simple definitions of the terms of $\mathfrak{J}$ (the special vocabulary of biology, or of psychology) in terms of $\mathfrak{R}$ (the special vocabularies of physics and chemistry, or of those and biology, say) are not forthcoming. Instead, what results are complicated strings of disjuncts and conjuncts that *either* are infinite or open-ended (and thus rule out a derivational reduction altogether), *or* are finite but (i) are too unwieldy to provide independent scientific value, or (ii) introduce disjunctions of $\mathfrak{R}$ which may not be projectable even though their disjuncts are.[5]

I fault the above concerns on two grounds. First, they overrate the relevance of derivational reductions (of any sort) to scientific practice itself; and, second, they mistakenly glue ontological concerns to issues of eliminability of vocabulary. At root the mistakes are the same: an obsession with words, which misses the real point: From the scientific point of view, *all* that is desired is an extension of the scope of an underlying science in a way illuminating both to that science and the special science above it.

Recall the examples from chemistry (§§ 2–3). Chemical kinds, as we've seen, *have been* reduced to physical kinds, and yet the deduction of every chemical law from physical law eludes us anyway. Crucial to the value of a derivational reduction is not the identification of vocabulary items but our capacity to extract suitable results from the laws of the underlying science. This is why, when it comes to derivational reductions, *scope* is the important notion, not program; and this is why suitably restricted derivational reductions, even when available, need not be particularly useful in explaining the interconnections between one science and another.

Consider a famous purported derivational reduction much discussed in the philosophical literature, the reduction of phenomenological thermo-dynamics

---

5   Kim (1992: 320) on the possibility of a disjunction (jade = jadeite or nephrite) failing the projectability test: "we can imagine ... on re-examining the records of past observations ... that all the positive instances [of 'Jade is green'] turn out to have been samples of jadeite, and none of nephrite!"

    Rosenberg (1985: 107) on the possibility of deductively reducing Mendelian genetics to molecular genetics:

> [This] founders on the impossibility of meeting the criterion of connection between the terms of the two theories. Such vast, unwieldy, general statements as we might construct – in which a Mendelian gene is equated with the molecular one – will be full of disjunctions, conjunctions, negations, exceptions, qualifications. It will make so many appeals to stages in the pathway between the DNA and the phenotypic endpoint of the pathway that it will be without any independent scientific standing. It will do no other work than substantiate a purely formal possibility without payoff for the actual advancement of either molecular or Mendelian genetics.

to statistical mechanics, something that involves what is often regarded as the reduction of the macro-properties of gases to statistical mechanics and the kinetic theory of matter.[6] We might initially describe the reduction in question as resulting in a *deduction* of certain macro-properties of a gas from its micro-properties (or, more accurately, from the properties and relations of the molecules making up the gases);[7] but this would be a mistake, for the *actual* micro-properties of the gases are not used in the derivation. Maxwell's original reduction relies on a number of idealizations, or, more bluntly, falsehoods.[8] In later, "more rigorous," reductions a number of these idealizations were removed. But a number were not, or were replaced by still other idealizations (e.g. about shapes of molecules, the nature of the forces interacting between them, the nature of the container walls, and so on).[9]

This reveals an important shortcoming of any analysis based on derivational reductions. Make the predicates general and vague enough ("aggregate of molecules," etc.), and an identification of macro-predicates with (truth-functional combinations of) micro-predicates that actually apply to something is possible; but nothing informative can be deduced from these identifications. On the other hand, introduce micro-predicates that are informative – because they are precise *and* mathematically tractable, say – and the result is that macro-predicates are identified with (truth-functional combinations of) micro-predicates which don't hold of anything real.

---

6   Many philosophers signal that this description needs qualification. Van Fraassen (1980: 10), for example, places "reduction" within scarequotes, and Kitcher (1984: 554) notes that "appropriate extra premises" may be called for.
7   According to Varney:

> Kinetic theory is a term applied to the study of gases by assuming that gases are composed of molecules subject to rapid random motions colliding with one another and with the container walls, analyzing these motions, and deducing various properties of the gases such as their molar heat capacities, the pressure they exert, their viscosity, coefficient of diffusion, and rate of effusion through apertures. The adjective *kinetic* refers to the fact that the properties of the gases are deduced from the *motions* of the molecules. The term *theory* refers to the *deduction* of the properties of the gases *without* reference to *any* known experimental properties, not even the ideal gas law (q.v.).

(1991: 601)

Van Fraassen (1980: 10) offers, parenthetically and in passing, the following identification of vocabulary items: "bodies of gas are identified as aggregates of molecules, temperature as mean kinetic energy, and so on."
8   For the original idealizations used by Maxwell, and some of his subsequent modifications, see Achinstein (1987: 242–3). "Falsehood" is perhaps too harsh. An idealization may be introduced by a practitioner without knowledge of whether or not it is true.
9   This use of idealizations cannot be removed *altogether*, for reasons already given: in order to carry out a reduction of this sort, the mathematics must be tractable; also we lack the technical tools to enable us to accurately describe the situation (e.g. the quantum-mechanical spread of molecules) without idealization.

The notion that derivational reductions informatively reveal the relation between special sciences and more general sciences stems from an uncritical adoption of theoretical deductivism. Given my discussion of scope limitations (§§ 2–3), and the standard methods used to circumvent such limitations, it should be no surprise that, in general, real derivational reductions of any sort are rare outside mathematics.

But, then, what *is* going on here? Well, suppose someone (Maxwell, say) thinks the various properties gases have are due to the statistical effects of the kinetic properties of their molecules. A derivation of these properties from a statistical description of the kinetic properties of the gas molecules would certainly show this. But if this cannot be done (the derivation is too complicated, the details of what the molecules of the gas are like are too messy …), something almost as good may be available: the derivation of these properties in an idealized context wherein certain complications are ignored. *Why* is this almost as good? Well it isn't *always* – idealizing *can* lead to trouble – but it is a good move when the complications involved can be (more or less) ignored because their impact on results falls below a certain threshold.

How do we tell if the impact of the ignored complications falls below a certain threshold? Broadly speaking, there are two ways. The first is the presence of (pretty good) empirical agreement between results predicted on the basis of the idealizations and what actually happens; this, of course, is a highly fallible consideration, for empirical agreement may be due to accidental factors. The second is if we subsequently see how to add in some of the complications previously ignored, crank through the derivation again, and get results rather close to what we got in the first place.[10]

Concentrating on the alleged ontological direction (from micro to macro), however, can make us overlook the real value of this sort of project, which is the other way around entirely. Macro-properties of gases, for example, are measurable; micro-properties of those same gases (e.g. average velocity of gas particles, mean length of the path of a molecule) are not. But one can use the measurable macro-properties to *infer* micro-properties, as Maxwell (1965) does explicitly.

Presuming micro-structure is not a simple matter of claiming that there are small particles bouncing around according to Newton's laws of motion. There are lots of details (e.g. about the nature of the forces between these particles) that are open to various options. But bringing facts about the macro-behavior of gases – in particular under what circumstances gases deviate from the macro-gas laws – together with general micro-facts about

---

10 This is what happened with the kinetic theory of gases, and it's what happens when the physical factors, being idealized away, really aren't contributing much. As Maxwell, and others, added complications (what might be called "more realistic assumptions"), they found the original results to be more or less robust.

particles, in particular Newtonian laws of motion, enabled Maxwell to infer much more specific facts about the statistical fine structure of gas particles. Thus, the *real* value of the project is not ontological but epistemological – to learn more about the micro-domain via the macro-domain.[11]

Two notions are linked together by this practice: there is, on the one hand, *explanation*, the explanation of certain macro-phenomena (e.g. why the macro-laws hold where they do, and why deviations from them occur when they do), and, on the other, there is *inference*, inference to the micro-facts of various sorts (about forces between molecules, etc.) that provide the explanations.[12]

In such cases, one should feel that what might legitimately be called a *reduction* has been carried out, despite the absence of anything like a derivational reduction. I call this a "scientific reduction," to contrast it with the sorts of reductions philosophers have been primarily concerned with.

Let's consider an objection the deductive holist has been wanting to raise for some time: the attack against the existence of derivational reductions (once again) relies on an overly narrow construal of scientific theories. The point of reduction is not to show that one narrow theory can be replaced by another, without the use of anything outside the scope of either. Rather, it is to show that whatever can be explained or expressed with the resources of the two theories (and whatever else) can be explained or expressed without the resources of one of these; it is to show that one of the narrow theories can be eliminated from the entire domain of discourse without explanatory and expressive loss.

Unfortunately for this objection, the important issue for the eliminability of discourse, when it comes to empirical science, is equivalence not of "explanatory power" or "expressibility" but of *applicability* – we, and this is the main lesson of indispensability arguments, cannot replace one theory

---

11 Maxwell (1965: 419) writes: "We have, in fact, to determine, from the observed external actions of an unseen piece of machinery, its internal construction." As an illustration of the kind of reasoning Maxwell engages in, consider this characteristic passage:

> But we know that most ordinary gases deviate from Boyle's law, especially at low temperatures and great densities. Let us see whether the hypothesis of forces between the particles, which we rejected when brought forward as the sole cause of gaseous pressure, may not be consistent with experiment when considered as the cause of this deviation from Boyle's law.

> (Ibid.: 423)

12 Harman (1965) dubbed this *inference to the best explanation*, and, contrary to how it appears in the hands of someone like Maxwell, irrealists such as van Fraassen (1980) explicitly deny that the explanation provided allows inference to anything true. I discuss this claim more fully in § 6. One point about Maxwell's procedure: he does not infer that gases have a molecular structure obeying Newton's laws of motion; he infers quite specific properties about that molecular structure on the basis of assumed general properties of micro-particles, plus details about the macro-behavior of gases.

with another unless we can *apply* the second *wherever* we can apply the first. Definitional derivations are strong enough to show that a theory can be eliminated in *this* sense only if the micro-predicates used in the identifications of micro- and macro-predicates *are the same ones* used to derive truths about the phenomena in question; only in this way will such derivations provide translations of sentences usable in *precisely* the same circumstances that the original sentences are used in. The way that idealizations and other auxiliary tools actually allow a scientific reduction thus need not offer a shred of hope of eliminating a domain of discourse.

Consider, again, the kinetic theory of gases. The point of scientific reduction in this case is to show that certain macro-properties of gases arise from statistical properties of the gas particles, and to infer details about the statistical properties of the gas particles from those macro-properties, *even in circumstances where we cannot describe these particles accurately*. But where we cannot accurately describe the micro-objects we cannot apply a micro-theory *directly to them*; and this means that the scope of the macro-theory outstrips the scope of the micro-theory. And so the macro-theory is *not* eliminable.[13] Indeed, the fact that Maxwell employed deviations from the macro-theory as evidence for further details about the statistical and dynamical properties of gas particles themselves illustrates this: we can apply the macro-theory easily to the gases (macro-properties, volume, mass, etc., are measurable), but this is not true of their micro-properties. What the scientific reduction allows is an *extension* of the scope of the micro-theory *via* the scope of the macro-theory already in place. But *this* scientific strategy rules out projects designed to *eliminate* macro-vocabulary.

The scientific reduction studied by Maxwell *is* compatible with a strict derivation of gas laws from: the kinetic laws of motion plus idealized descriptions of the molecules of such gases plus premises, empirically established in the way I've described above, that the idealizations introduced do not ignore significant aspects of molecular behavior as it manifests itself in the interrelation of macro-properties of the gases. But this derivation is not a derivational reduction because, I repeat, it cannot match the ranges of the macro-predicates with the ranges of the micro-predicates; and it can't do this because it doesn't supply a vocabulary – micro-predicates – that we can *apply* in (most, if not all) real situations.

Nevertheless talk of there being a *scientific reduction* in this sort of situation is still legitimate because we really do take *A*s, and what is going on

---

13 This is not a problem about the expressive power of the micro-vocabulary (imagine it to be as rich as you like); this is about whether, and under what circumstances, we can recognize that a certain micro-vocabulary *applies to* certain phenomena. Also, replacements of macro-vocabulary by broad and vague micro-predicates ("aggregate of molecules") that *are* usable *is* possible, but pointless.

with them, to be nothing more than *B*s, and what is going on with them; we recognize and expect that if, in certain cases, we overcome (particular) tractability problems (as we sometimes do) in treating *A*s as *B*s, we will not discover recalcitrant emergent phenomena.[14] Scientific reduction is a project with methodological *depth*: the idealized model is one where deviations from what is actually going on are deviations we can study directly, extract information from, and, when we're lucky, minimize. This is the full content of the claim that *A*s, and what is going on with them, are really just *B*s, and what is going on with them.

So the metaphysical claim (*A*s are just *B*s) is not one established through definitions of *A*s in terms of *B*s, or anything like that. It emerges in the practice of scientists in *treating A*s as *B*s whenever tractability permits (if a wingless rodent falls out of an airplane, we'll treat it as a falling body, and not worry about the possibility of emergent aerodynamical phenomena because it is a *rodent*), and in the use of our epistemic access to *A*s to infer facts about otherwise epistemically inaccessible *B*s. This metaphysical belief shows itself also in the clear-headed recognition of where reality leaves off and idealization (most likely) begins, which gives the scientist space to tinker with the models, if tinkering is necessary.

Let me summarize what I have claimed here. What does it mean to be a physicalist? You might have thought it meant that everything you want to say that's true is something you should be able to say using just the vocabulary of physics. You might also have thought that it meant you were only committed to physical things, that talk in any other terms was talk of the unreal.

I've suggested that a commitment to physicalism need have nothing at all to do with issues about vocabulary, and, consequently, nothing at all to do with ontological reduction, as philosophers understand that idea. Rather, a commitment to physicalism can be identified solely by the adoption of a certain methodology: one treats failures to successfully apply physical laws to a domain as due to intractability, and not to the simple inapplicability of the laws to that domain. Consequently, one studies the domain in question and the tools available with the aim of reducing this intractability to some degree, in the expectation of sharpening the results to the extent that tractability increases. That's *all* a commitment to physicalism requires, and all that's needed.

I make analogous claims about other nonreductive[15] doctrines. If one

---

14 See the objection in what follows to token–token physicalism for an example of "recalcitrant emergent phenomena."

15 Where by "nonreductive" I mean pretty much what Kim (1989) means: a doctrine unaccompanied by either the elimination of offending vocabulary or by anything like a derivational reduction of such vocabulary to underlying vocabulary.

thinks that psychological events are entirely neurophysiological, what's required is not a linkage, however tenuous, between the vocabularies of psychology and neurophysiology, but simply the methodological intention to laboriously extend the scope of neurophysiology to psychological events, to use our epistemic access to psychological events to infer facts about neurological stuff, and to realize that success in these endeavors is directly proportional to one's escape from intractability.

One philosopher seems to have argued similarly. Fodor (1974: 138) writes:

> It seems to me ... that the classical construal of the unity of science has really badly misconstrued the *goal* of scientific reduction. The point of reduction is *not* primarily to find some natural kind predicate of physics coextensive with each kind predicate of a special science. It is, rather, to explicate the physical mechanisms whereby events conform to the laws of the special sciences.

Fodor offers token–token physicalism, the doctrine that any event (under any description) is a physical event (of one sort or another) as an adequate and complete construal of physicalism. This view certainly detaches physicalism from the requirement of supplying (suitably restricted) derivational reductions.[16]

Unfortunately, token–token physicalism is too weak to suitably characterize physicalism because it can be satisfied in circumstances where we would deny physicalism was vindicated. This can be easily seen by reverting to the case of the reduction of phenomenological thermodynamics to statistical mechanics. Imagine, contrary to fact, that a genuine derivational reduction is available, but only if constraints are placed on gas states that are – given the physics of micro-particles – quite probabilistically low. In such a case the situation *is* one that satisfies token–token physicalism (each macro-event is one or another physical event); but since the admissible physical events in question are of extremely low probability (physicalistically speaking), physicalism fails: emergent phenomena, indicated by physically inexplicable constraints on the probability space of micro-particles, show this.[17]

---

16 The vocabulary must still be definitionally eliminable in terms of truth-functional combinations of physical terms, however, or the token–token physicalist must reject Quine's way of recognizing ontological commitments. Kim objects – as we've seen – that such derivations don't preserve projectability, and aims this point at Fodor's token–token physicalism. I do not address *this* objection to token–token physicalism, which I find artificial; I raise another.

17 Garfinkel (1981: 70–1) seems aware of this possibility, but seems also, falsely, to think that the actual derivation of the Boyle–Charles law from the statistical behavior of the ensemble of molecules illustrates it just because of the use of the conservation of energy (in closed systems) and the assumption of a normal distribution of velocities.

To summarize: I've offered a characterization of physicalism which denies that vocabulary should dictate ontological commitments. And this means a *rejection of* both the Quinean criterion for ontological commitment and the related use of vocabulary criteria for ontological programs.[18]

I conclude this section with an illustration of a thesis rejected by my sort of physicalist. Kim (1992: 322) characterizes "the Physical Realization Thesis" as having two clauses: namely, that

1 "pain, or any other mental state, occurs in a system when, and only when, appropriate physical conditions are present in the system"; and
2 "significant properties of mental states, in particular nomic relationships amongst them, are due to, and explainable in terms of, the properties and causal-nomic connections among their physical 'substrates'."

Kim observes that this doctrine is widely accepted by philosophers, especially functionalists.

My scientific physicalist objects to the second clause: however "explainable" is understood by the proponent of the physical realization thesis, it is only under special and rare circumstances that significant properties of mental states can be *explained* in physicalistic terms. Furthermore, the scientific realist doubts "explainable" can be treated metaphysically or dispositionally, for this presupposes some idealized way to overcome tractability problems, and there is little reason to believe in such a thing.[19]

18 I am sounding a theme that will loom large later on: a kind of split between what we *say*, the vocabulary we are forced to have, and what we *do*, or how we treat what that vocabulary apparently refers to. See Part IV.

    I have taken myself, in this section, not to be attacking traditional approaches to coding metaphysical attitudes in terms of constraints on vocabulary, but to be offering an alternative notion of physicalism more in accord with actual scientific practice. There is a great deal left to say about why, in particular, the Quinean approach to ontology is problematical. See Azzouni (1997a,b and 1998) for details.

19 Dupré (1993: 95) notes that some philosophers have distinguished between the practical and the theoretical possibility of a reduction: "The paradigmatic conception of a practical obstacle is mathematical or computational complexity." I do not think such a distinction can be taken seriously. In describing as *epistemic* the mathematical problems facing attempts to apply a mathematically articulated science to a domain, I was not suggesting that it makes sense to claim there is some way, *in principle*, to so apply a science. This suggestion is troubled in two ways: first, it underrates what mathematical intractability comes to; and, second (by implicitly subscribing to theoretical deductivism), it fails to see the rich and complicated ways we standardly employ to get around intractability, ways that have little to do with the reduction of one science to another.

§5

SOME COMMENTS ON THE PHILOSOPHICAL IMPLICATIONS OF THE USE OF IDEALIZATIONS IN SCIENCE

The failure of theoretical deductivism and the muting of confirmation holism are purely epistemic both in character and in their implications. Nancy Cartwright (1983), making many of the same points about scientific methodology that I've raised in § 2, has drawn metaphysical anti-realist conclusions from them, specifically the denial of the truth of high-level physical laws. I object to this line of thought in what follows, as it arises in her work and in that of others: I argue that the truth-predicate must play a special role in scientific theories, and that this constraint is not respected by the kind of anti-realist strategy she and others employ.

Cartwright's first move (following van Fraassen and Duhem) is to separate explanatory adequacy from truth, and to add an additional requirement on taking something to be true over and above its capacity to successfully explain. The idea is this (1983: 87): an explanation for something, an explanation that uses tools (theoretical laws) which "organize, briefly and efficiently, the unwieldy, and perhaps unlearnable, mass of highly detailed knowledge that we have of the phenomena," is certainly invaluable, but "organizing power has nothing to do with truth." She offers Ptolemaic astronomy as an illustration of a theory which explains the motions of the planets but which isn't taken to be true (even by its practitioners).[1]

---

1 I distinguish between something *being* true and something *being taken* to be true. Cartwright is concerned with whether we should *take* the theoretical laws of physics to be true, even though she sometimes writes as if she is concerned with whether they *are* true. Truth *is* of concern, of course, but only because of a prior concern with whether purported reasons for adopting physical laws, e.g. inference to the best explanation, are sufficient for *taking* them to be true. (It should be obvious that *any* reason we give to justify a claim that a scientific law is true might fail in its force; this is why taking-as-true is what is at issue.)

On the historical veracity of her claim about Ptolemaic astronomy, recall footnote 27 to § 2. For purposes of discussion, I continue to treat the example as do Cartwright, van Fraassen, Duhem, and others.

What else is required for theoretical statements to be true? Cartwright puts this question in the form of a challenge drawn from van Fraassen (1980): "to give an account of explanation that shows *why* the success of the explanation, coupled with the truth of the explanandum, argues for the truth of the explanans." She imagines two ways to meet this challenge. The first, which she rightly regards as implausible, would be "[i]f we could imagine that our explanatory laws *made* the phenomenological laws true ..."[2] The second, the generic–specific approach, taken from Grünbaum (1954), is to treat theoretical laws as more general cases of the specific phenomenological laws often applied to particular situations.[3]

The burden of much of Cartwright's 1983 book is to show that the second way of meeting van Fraassen's challenge fails. It fails because the deductive relation required by the generic–specific approach between theoretical laws and phenomenological laws is not available.[4] Thus, the idealizations, short-cuts, empirical adjustments, and so on, that physicists use to move from theoretical laws to phenomenological laws indicate that the repeated confirmations of the phenomenological laws yielded by their successful application in physical situations do not trickle up to the theoretical laws themselves.[5]

Cartwright's strategy for denying the truth of theoretical generalizations in physics is vitiated by a failure to appreciate the role of the *truth-predicate* in our web of belief. To see this, let's first rehearse the debate in philosophy of mathematics over the Quine–Putnam indispensability thesis.[6] The idea is

---

2   Cartwright (1983: 94).
3   Theoretical laws are high-level generalizations taken to be explanatory; phenomenological laws, by contrast, are descriptive. See the Introduction to Cartwright (1983) for a discussion of how she understands physicists to use the distinction. However, she goes beyond strict physical practice when she argues that phenomenological laws are true in contrast to theoretical laws, which are not. I'll not consider this distinction much further except in one respect: Cartwright seems to treat all theoretical laws as alike. But it will become clear that only *some* of these are seen by physicists as fundamental, in the sense of being taken to be true (for now) without qualification or approximation.
4   Cartwright (1983: 107): "The generic–specific account fails because the content of the phenomenological laws we derive is not contained in the fundamental laws which explain them."
5   Cartwright writes:

> If we are going to argue from the success of theory to the truth of theoretical laws, we had better have a large number and a wide variety of cases. A handful of careful experiments will not do; what leads to conviction is the widespread application of theory, the application to lasers, and to transistors, and to tens of thousands of other real devices. Realists need these examples, application after application, to make their case. But these examples do not have the right structure to support [the realist thesis. For the laws do not literally apply to them.

(1983: 161)

6   For references and further details, see Azzouni (1997b).

that the application of mathematical discourse to empirical science leads to a commitment to the *abstracta* presupposed by the mathematics so applied, because the mathematical theorems, and, more importantly, the mathematical frameworks that empirical theories are couched in,[7] are indispensable to the *formulation* as well as the deployment of scientific (specifically, physical) theories.

The argument involves two steps: first, there is a commitment to the *truth* of that mathematics used in empirical theories; second, a Quinean criterion of ontological commitment secures a commitment to the *ontology* of the mathematical discourse thus taken to be true.

Let's concentrate on the first step. What secures commitment to the *truth* of the mathematics used in scientific applications? The literature is not clear on this point: somehow need for *application* creates commitment in some way to the *truth* of what is indispensably applied. In fact, what does the job here is not only the *mathematics* used, but the indispensability of the truth-idiom *itself* to the indispensable amalgam of scientific and mathematical lore. That is, what creates a commitment to the truth of empirically applied mathematics is the need for a truth-predicate in such applications, a predicate that traditionally takes Tarskian form.[8]

The point is made explicit by Quine (1970b): asserting a sentence, or a finite set of sentences, can be done directly, by exhibiting the sentence(s). But to assert a class of sentences that is *not* finite, or is not accessible except through description (e.g. the first sentence asserted in the year 3000), or to describe the consequences or effects of a class of sentences without mentioning them explicitly, we need a device that amounts to quantification over sentences. That such a device is indispensable to empirical application, and indeed, to our web of belief generally, is easily recognized by the fact, *among many others*, that number theory is not finitely axiomatizable. Unwieldy collections of sentences arise often in empirical science, and so the truth-predicate is indispensable to the drawing of consequences of all sorts from scientific theories.[9]

The indispensability of the idiom "true" as a crucial premiss in indispensability arguments can be seen by imagining a case, contrary to fact, where all

---

7 E.g. the mathematical formalism for quantum mechanics.

8 I speak of a truth-*predicate* for concreteness; use of other devices to the same purpose – operators, sentential substitutional quantifiers, and so on – won't affect the argument.

9 What might well be called the quantificatory role of the truth-predicate has been noticed by many: Leeds (1978); Putnam (1978b); Quine (1970b); and so on. There is a lively debate as to whether this is its *sole* role, but that particular issue may be left aside for the moment. I will say a few things about truth that bear on that topic subsequently, and in Part II § 7. I should add that I have already (Azzouni 1994) made a big deal of the idea that truth – restricted to this role – suffices for mathematics. Resnik (1990 and 1997) makes similar claims for "immanent truth."

applications of scientific and/or mathematical lore requires at most the explicit exhibiting of sentences. In such a case, nothing would prevent us from classifying *individual* sentences as we wished – *these* are true; *these others* are only instrumental – especially if mathematical statements (and physical generalizations of a certain sort, say) were never themselves used in the application situation but only mediately applied via other (observation) sentences or phenomenological generalizations. The difference in the practical use of individual sentences, that is, could itself mark the difference between true sentences and those sentences instrumentally required for applications of our web of belief, but which are false. But once a truth-predicate is required to operate quantificationally over classes of sentences, this strategy is ruled out.

Essential to a truth-predicate, if it is to do the job described above, is that it be topic-neutral:[10] It must apply to (or range over) *any* sentence of a conceptual scheme that we use in empirical applications, specifically if implications are drawn from that sentence along with its fellows. Thus sentences indispensable to the application of the web of belief cannot be singled out as false (or *other* than true) by the truth-predicate if it is to do the job it is designed to do.[11]

The topic-neutrality of truth places a *very* heavy burden on fictionalists and their fellow-travelers, in particular:

(i) those philosophers who single out a particular class of sentences ordinarily used in our conceptual scheme (mathematical sentences, say, or high-level laws of physics), and describe them as false (or merely as "true in a story"); and

(ii) those philosophers who add requirements to the notion of truth – over and above its quantificational role – which have the effect of distinguishing as true some sentences in our web of belief apart from others, which are not true but whose role in the conceptual scheme is otherwise equally indispensable. For these philosophers must undertake the arduous task of showing that the set of sentences currently crucial to our conceptual scheme, but which on their view is "fictionalized," are in fact *dispensable* because otherwise the topic-neutrality of truth is violated.

Consider Field's 1989 fictionalism, directed towards mathematical sentences

---

10 The truth-predicate shares this property with logical idioms generally, and for related reasons. See Azzouni (1994) for a discussion of the topic-neutrality of logical idioms.

The topic-neutrality of truth must be restricted in some way if self-reference is involved, as the literature on liar paradoxes shows. I leave this issue aside.

11 I explore ways of qualifying this constraint presently.

committed to *abstracta*: fictionalism stands or falls with a substantial program to show that such mathematical statements are dispensable, namely, his efforts at nominalizing physics.[12]

Since the topic-neutrality of truth disables any attack on the Quine–Putnam indispensability thesis, which faults the *truth* of mathematical claims without accompanying that claim with a *successful* Field-style project of showing the dispensability of such sentences, the only route still available for opposition to the indispensability thesis is that of attacking the Quinean criterion of ontological commitment instead, claiming that mathematical statements, indispensable in application, *although true*, do not carry the ontological freight they seem to.[13]

Let's return to Cartwright. I've already noted her claim that if a scientific law is to be taken as true, more is called for than simple explanatory success, together with the concomitant virtues of neatness, organization, predictive power, and so on, since when these are possessed by theories like the Ptolemaic theory, this still isn't sufficient for us to take them as true. Unfortunately, Cartwright's allusion to the practical indispensability of theoretical laws of physics ("organize, briefly and efficiently, the unwieldy, and perhaps unlearnable, mass of highly detailed knowledge that we have of the phenomena") opens her to the objection that she has placed a constraint on truth that violates its topic-neutrality. Indeed, it is very unlikely that high-level generalizations in physics *are* dispensable altogether from empirical application, if only because of the role she recognizes they have, and so her view falls foul of the topic-neutrality of the truth-idiom on these grounds alone.[14]

It is worth taking a moment to wonder whether van Fraassen's 1980 "constructive empiricism" also overlooks the topic-neutrality of truth. It seems not to, but the matter is delicate. Van Fraassen takes good scientific theories to be not true but only empirically adequate – where that means the

---

12 Arnold Koslow has pressed me for textual evidence that Field recognizes the dependence of his fictionalism on his nominalist program, and I haven't found any; so my claim about his views may be regarded as a charitable one. One philosopher who fails to recognize the pertinence of the topic-neutrality of the truth-predicate to fictionalist views of mathematics is Mark Balaguer (1998: Chapter 7 § 3): he offers the fictionalist a notion of "nominalistic content" without any hope of distinguishing nominalistically acceptable *sentences* from other sentences and showing that only the former are indispensable to applications of mathematics. Consequently, the Balaguerian fictionalist must abjure use of the truth-predicate altogether, because all sentences of the conceptual scheme turn out false on his view. The pragmatic uselessness of the web of belief then follows. See the discussion of van Fraassen's "constructive empiricism," later in this section, where similar objections are raised about a notion of observational content.

13 See Azzouni (1997a and b, 1998), where I explore these options.

14 I show later in this section that such laws play an additional evidential role that gives reason for their indispensability beyond the pragmatic reasons Cartwright considers.

52

theories are right about the observable. Since van Fraassen also accepts that our language is theory laden – he seems committed to the result that our entire web of belief is *false* (for all we know). Having repudiated (1980: 55–6, and elsewhere) the possibility of purely observational sentences, he can't be accused of failing to notice the topic-neutrality of truth. But his position does seem to have the very odd consequence, just noticed, of our being unable to assert even one truth; and, therefore, it seems to rule out altogether using a truth-predicate on our sentences.[15]

Perhaps we don't have to saddle van Fraassen with the above position. For he also thinks an observation–theory distinction can be drawn within science itself, and that this distinction is one about *entities*, not terminology. This opens the possibility of finding a class of sentences – about such specific entities or classes of entities – that can be regarded as true – although, because of the theory-laden character of all our language, and because of the topic-neutrality of truth, he will then be forced to the acceptance of the truth of all the sentences in our web of belief.[16] Still possible, though, is salvaging a portion of van Fraassen's anti-realism by adopting a move I've mentioned before: denying that existential quantifiers carry ontological weight. This would involve using the observation–theory distinction as applied to *entities* to demarcate those that exist from those that don't, even while accepting the truth of *all* the sentences in our web of belief, both those about entities that exist and those about the others.[17]

15 Van Fraassen says (1980: 58) that theory-ladenness "does not obliterate the distinction between what is observable and what is not – for that is an empirical distinction – and it does *not* mean that a theory could not be right about the observable without being right about everything." He seems, therefore, to be gesturing towards a notion of observational content without allowing that we could express such content (exclusively) in sentences. Is this position utterly untenable, as I suggested (note 12, above) of Balaguer's analogous position? Van Fraassen (1980: 12–13, 14, 91–2, 202) offers some provocative remarks on how scientists can adopt a language or theory without committing themselves to its truth. In particular he writes: "We have to make room for the epistemological position ... that a rational person never assigns personal probability 1 to any proposition except a tautology" (p. 9). I wonder whether this view can be squared with the quantificatory role of the truth-predicate: van Fraassen's view, as it is expressed here, seems to lead to the conclusion that we must take such an attitude to everything we say, excluding tautologies, but *including* non-tautologous remarks about the *truth* of our statements. Perhaps this conclusion can be avoided by a refusal to identify use of the truth-predicate in respect of a sentence with assigning that sentence a probability of 1. Is *this* cogent? Maybe, but I won't explore this avenue further now, since it is not the view van Fraassen takes.

16 Or he can adopt a non-standard notion of implication, so that true sentences can have false consequences.

17 This move is compatible with much of his text, although, of course, not with his remarks on truth. He writes (1980: 81): "[I]mmersion in the theoretical world-picture does not preclude 'bracketing' its ontological implications." One can do this while *still* taking such

The indispensability argument, as I've presented it, seems to show too much: Cartwright's claim, that those using Ptolemaic astronomy could still reasonably deny its truth, sounds right. One way the proponent of the indispensability argument can accommodate this point is to nuance the requirements of indispensability: a part of our web of belief – a specific theory, say – can be treated instrumentally if it can be isolated from the rest of our web in certain respects.

Ptolemaic theory, for example, could be treated as an entirely instrumental tool for predicting the locations of the planets, even by practitioners without an alternative astronomical *theory*, because they had other means of recognizing planetary locations (i.e. observation by the naked eye); and the knowledge brought to bear to recognize planetary locations this way is not confirmationally dependent in any way on Ptolemaic theory itself. Also, and this is equally important, there is a place to stand, metaphorically speaking, from which practitioners of Ptolemaic theory can describe Ptolemaic theory itself, its data, and its predictions; and this (linguistic) standpoint, too, is not confirmationally tainted by Ptolemaic theory.

At the moment, however, I'm less interested in how scientific theories can be isolated in the appropriate way from the rest of our web of belief so that they can be treated instrumentally than in how the application of (certain) scientific laws can force them to be laws we *cannot* treat merely instrumentally, but must take to be true (at least until we replace them outright with something else).[18]

Let's consider how a law can, in general, prove indispensable in empirical applications. One way, of course, is if it applies directly to experience without the benefit of idealizations and the other mediators explored in § 2. Another way is if the law is a (psychologically indispensable) codifier of more specific facts which are too scattered for us to retain directly. Cartwright, as we've seen, recognizes both these ways.

Is there a third, a way that she overlooks? Yes: if the law, taken exactly, is used in evidential arguments about the domain it applies to. The law itself in such cases helps classify the status of empirically measured deviations from it; how we classify, for example, exactly what the idealizations (and the other mediators of § 2) are doing turns crucially on taking the law (that the idealizations and other mediators are introduced in respect of) to hold exactly.

A law, used *this way*, is evidentially indispensable to our web of belief, with, of course, the rider that pressing the law as exactly holding in a

---

theories to be *true*. I take the marking out of observational entities to require terms with only what I call perceptual procedures. See Part II § 3 on such terms and their properties.

I should add that van Fraassen's focus on the semantic construal of scientific theories as opposed to the syntactic construal will not enable him to elude the issues about truth I have raised here – the same issues arise at the meta-level.

18 I return to the method of isolation towards the end of this section.

domain can lead to its being overthrown. In that eventuality, we may have to reclassify the deviations from the new law in new ways. But *until* the law is so overthrown, it plays a crucial role in characterizing the nature of empirical phenomena – in particular, empirical deviations from it.

A simplified example will illustrate what is involved here. Imagine that some empirically measurable phenomena are a function of (among other things) the geometry of an otherwise inaccessible object. If we take the mathematical description of how the phenomena are a function of the object's shape *as exactly true*, and if the geometry of the object is represented in the function describing the phenomena in a way that links various shapes so they can be seen as approximating each other, we can start with a guess at the object's shape, and use deviations in our predicted results from empirical measured results and the mathematical description of the function to yield a more accurate description of its shape. In this way the theory itself is used to show what the object's shape must be, by introducing what George Smith has described as a series of successive approximations.[19]

Three points. First, the way that theory, in these cases, is essential to a characterization of empirical deviations from the theory is one of the things that led such philosophers of science as Kuhn and Feyerabend to propound incommensurability. For the thought is that data tainted *to this extent* by theory cannot supply an independent standard by which to compare competing theories. This is an issue I discuss later; my point now is that such uses of theory are neither uncontroversial nor unnoticed, despite the result that such theories, as long as they are held, are profoundly indispensable to empirical application.

The second point is that one may suggest, on behalf of Cartwright, that the truth-operator is indispensable to the empirical application of a set of sentences only if we explicitly *deduce* consequences from them, and not if we use them merely as a framework within which mediators to empirical applications can be introduced; in this way, the quantificational role of the truth-predicate might, one hopes, be shown to be unnecessary for the theoretical laws Cartwright wants to regard as false. This can't work, if only

---

19 Smith (forthcoming). Smith illustrates Newton's systematic use of precisely this methodology in his *Principia*, with respect to 3/2 power rule and planetary orbits. Smith also offers a useful list of "categories" of deviation, and shows how a scientific theory (taken exactly) helps in the classification of their statuses (as observational errors, deductions incompletely carried through, unnoticed gravitational forces, unnoticed non-gravitational forces, or, last but hardly least, deviations leading to an overthrow of the theory itself).

George Smith has also drawn my attention to another example of the method of "successive approximations": the previously cited "On the Dynamical Evidence of the Molecular Constitution of Bodies," in which Maxwell uses empirically measured deviations to provide evidence about the forces between molecules. There are, I need hardly say, numerous examples of these methods in contemporary physics.

because (a) the classification of deviations that a theory helps to induce is one that still involves deductions of various sorts; and (b) in describing different approximations, and why they should yield empirical results of diverse kinds, semantic ascent – the explicit discussion of what is compatible with a theory – is required, and this makes the truth-predicate, if anything, even less dispensable.

Third, it is important to realize that the method of "successive approximations," as Smith describes it, is *not* what Boyd (1981: 613) means when *he* writes of "successive approximation." Boyd means to suggest that (almost all) of our current scientific claims are at best only approximately correct, and that, moreover, "the notion of *exact* truth plays no significant role in the realistic account of the reliability of scientific methodology. The reliability of the scientific method does not depend on the exact truth of background theories ..." (ibid.: 630). I understand Smith's successive approximations to turn crucially on taking (at least some) scientific laws as *applying exactly*, and as using empirically measured deviations from scientific predictions evidentially. I will argue presently that Boyd's view, like van Fraassen's and others', cannot be squared with the role of the truth-predicate; it should already be clear that it will have a hard time explaining the evidential practices just mentioned, which turn on taking (certain) scientific laws as exact.

I'm clearly committed to a distinction between theoretical laws taken as true *simpliciter* and those taken as approximating truth. Nowhere in her 1983 discussion of the truth-conditional status of the theoretical laws of physics does Cartwright distinguish between these, apparently, like Boyd, treating all theoretical laws alike insofar as realist attitudes are concerned. But this distinction is important because only laws taken exactly are required to be true; approximating laws are recognized as approximating with regard to other laws taken exactly.

Let me give some elementary examples of laws of each type. Laws currently taken to be "true" *simpliciter* are the equations of general relativity; at an earlier time, Newton's laws of motion and his force law for gravity were taken as exact. When results are derived from such laws, or, far more commonly, when approximation methods are employed to yield results we take to follow from such laws, *either* those methods are specifically understood to be there for reasons of tractability, and one tries to refine the "derivations" to see if they still hold under more realistic assumptions, *or* the approximation methods are there to provide evidence in the form of successive approximations.

On the other hand, "loose generalizations" are not taken to be true *simpliciter*. They are seen as generalizations that hold (except for small deviations), provided certain things are small enough, slow enough, fast enough, etc.[20] When mediating tricks are employed to derive *these* from the fundamental laws we take to be true, it's recognized that suppressed factors allow results close enough to the real values to satisfy our needs.[21]

Furthermore, when *additional* approximation methods provide a yet better fit between loose generalizations and empirical yield, often the approximations introduced *really do* provide something we can regard as closer to the truth than the original loose generalization; this is because such additional approximations may take account of factors not quite small enough, slow enough, whatever, for the loose generalization in question to predict sufficiently accurate empirical results. This is part of what's involved in classifying empirical deviations from the law: *they* can be corrected or approximated.

Despite the language used, which might suggest otherwise, I don't think that loose generalizations possess something called "approximate truth." Loose generalizations are *false simpliciter*, and so in a sense talk of "approximating truths" misses the point entirely (nevertheless, I'll continue to use the idiom). Sentences are either true or false, and that is as far as *truth* is involved here. Loose generalizations are valuable because of their content: the way they describe things enables us to produce numerical answers within a certain specified neighborhood of the correct answer.

Such answers *can* be the result of the application of loose generalizations because, for example, something which is falsely described as a sphere can be rather close to a sphere in shape. But no such qualitative similarity in what is described is *required* in order for a loose generalization to work. If we can crank out answers numerically close to the right ones by describing something in a way that is flagrantly wrong (we describe something as filled with a fluid when it is not, or we describe a piece of matter as if it is a coil of wire when it manifestly is nothing like that at all), that's acceptable, too. The important point is to be aware of what we're doing, and why.

I have also described loose generalizations in a way that makes them "semantically structured." It may seem, therefore, that we can analyze a loose generalization into, say, a conjunction of sentences in which the true "respects" of the loose generalization are separated – by ampersands – from its false "respects." This is sometimes true, and is often explicitly recognized by unanalyzed residual terms in equations (which are neglected in calculations, except in special circumstances). But, in general, this can't be done explicitly – that's why loose generalizations are used in the first place.

---

20 Recall that I noted the presence of these generalizations in § 2, under the heading of "calculational shortcuts."
21 Historically speaking, a law is often downgraded from being something we regard as true *simpliciter* to a loose generalization. This is what happened to Newton's laws of motion, and there are many other examples.

Finally, consider the following: I have described examples where loose generalizations are "derived" from fundamental laws, and indeed this seems to me to be how things are in contemporary physics; but it is logically possible to have loose generalizations without being able to write down the fundamental *physical* laws that are, as it were, behind the generalizations. Imagine that the force laws for several forces are not as force laws currently are, independent of each other, but instead that the form taken by the law for each force depends on the strength of other forces in the universe. If this happened, a "second-order" version of *n*-body problems could arise for *laws*: although we might be able to prove, mathematically speaking, that physical laws of a certain sort exist, we would not be able to write down the actual laws but only numerical approximations of them which, mathematically speaking, we could sharpen *ad infinitum*. This is, however, an entirely imaginary case, as far as I know.

Here is a worry that the foregoing analysis might raise: I've argued that approximating laws should not be regarded either as true or as even approximately true. And they do not have the indispensable evidential role of laws that are taken to be exactly true. Nevertheless, they may still be empirically indispensable – in the second sense of being codifiers of specific descriptions, or in the sense that they are needed to be applied against experience because the laws we take to be exactly true cannot be directly applied. But, then (the worry goes), don't the considerations raised about the topic-neutrality of the truth-predicate require us to take them to be true?

No. We can classify them in respect of a background theory (laws taken exactly) which we do regard as true, and we can show that such laws yield acceptable approximate results because of various factors (e.g. deviations falling beneath certain thresholds). This meta-linguistic framework, where we study the empirical applicability of approximating laws against a background of laws taken as true *simpliciter*, is confirmationally isolated from the approximate laws themselves, much as with the Ptolemaic case. Furthermore, the pragmatic requirement of codification is *not* handled by approximating laws, but by those same (exact) laws used as a background to justify approximating laws in their domain of application.

This answer also handles the thought experiment about *n*-force law problems. Here the worry is that any law that we could write down would have to be regarded as false, and so what background context (of true theories) would be available to enable us to evaluate the approximating character of such laws? Well, the context would have to be an officially meta-linguistic one – one in which the principles used to generate the approximating laws themselves would be what we take to be true.[22]

---

22 That is, this would be a situation where meta-mathematics would be the context required to describe physical laws.

So here is how the situation shapes up. We have, first, a distinction between phenomenological laws and theoretical laws. The former, being descriptive, are precisely the laws that need direct and repeated confirmation: for descriptions are supposed to guide us when we're, as it were, pressed right up against the tribunal of experience. But theoretical laws do not require this. Rather, their relationship to experience is less direct, and is hampered by the presence of diverse sorts of intractability. Thus they are brought to experience through the medium of different sorts of idealization. But this does not allow us to think them false, as Cartwright urges, unless they are either dispensable, or isolatable from the rest of our web of belief (which we take to be true). Theoretical laws, thus, divide nicely into those which are fundamental (true *simpliciter*) and those which aren't (loose generalizations). They do anyway, *when we know enough about them*. If a theoretical law is taken to be fundamental, idealizations and approximation methods used to apply it are not taken to improve it: Rather they are there solely for purposes of tractability. On the other hand, when a theoretical law is not fundamental, such idealizations and approximations may very well improve on it.[23]

One caveat. A law is not approximate just because it leaves something out. Consider a gravitational-force law which does not characterize the effects due to other sorts of force (electromagnetic forces, strong forces, and so on). Such a law is not taken as approximate, even if there is nothing the law applies to all by itself.

Why the contrast with, say, Newtonian laws of motion? The reason, I think, is that the relativistic laws of motion are mathematically neat formulas which supersede Newtonian laws in their own proper domain of application. This is not the case when laws for new forces are added to the physicist's toolkit of previous force laws. Rather, since new forces (so far) do not affect the presence or absence of old forces, the laws for the forces are independent of each other (they are "additive").[24]

---

23 As an illustration of a case where we don't know enough about a generalization to classify it, consider Cartwright's discussion (1983: 113–18) of the derivation of the exponential decay law from the Schrödinger equation. She discusses two treatments, the Weisskopf–Wigner method and Markov treatment, in fair detail. In both cases, approximation methods are employed, which indicates that no strict derivation of this law from the Schrödinger equation exists. Well, which sort of law is it? For example, is decay really exponential, or is the law in question a loose generalization? There seems to be no empirical evidence (yet) to decide the issue one way or the other, and the approximation methods used in the "derivations" of the law do not seem sufficiently well understood to tell from *them* what is going on.

By the way, Cartwright knows all this, and, indeed, says as much (1983: 118). So why on earth does she take this to be an illustration of an approximation "that improve[s] on laws"?

24 Another issue philosophers sometimes raise is the worry that, for example, the force law for gravity is only vacuously true because the presence of other forces shows there is nothing in

Another observation. General relativity breaks down in the context of singularities (in black holes; at the instant of the big bang), and clearly physicists take the existence of such breakdowns to point beyond general relativity. But, despite this, the laws of general relativity cannot be taken as approximate *now* because there is yet no place to view the laws of general relativity *from*. This is an extremely important consideration. It is quite common – this is what, I think, motivates Boyd (1981) to regard scientific laws *in toto* as approximate – to engage in an inductive generalization on the history of science that *no current* scientific truths are true *simpliciter*. But just as we cannot stand back from general relativity *now*, and claim that it is false (for lack of contrast with something we can describe that's *true*); so, too, we can't attempt to speak from the context of future science and announce that all of our current science is only approximate. Approximations can be described from a context exactly understood and against which they can be measured; otherwise claims about approximation have no content. It's commonly pointed out that we cannot stand outside our own conceptual scheme, but it's overlooked that views about the falsity or approximate nature of currently indispensable truths try to do exactly that.[25]

There are two loose ends I should wrap up before offering a brief summary. First, although I have argued that consideration of the indispensability of the truth-predicate's role commits us to the truth of the indispensable laws of our web of belief, there are many philosophers who, aware of the role of the truth-predicate, still think themselves committed to a minimalist theory of truth – one which, presumably, cannot be used to support scientific realism the way I do: at least it can be argued that I have secured the "truth" of scientific law in only a very weak sense.[26]

Minimalist claims that truth can be understood in a metaphysically deflated way must be presented very carefully so that they, too, do not violate the topic-neutrality of the truth-idiom. This is because, as we've seen, any condition on "truth" to strengthen it into something "metaphysically significant" which, at the same time, separates some sentences in our conceptual scheme from others that are otherwise indispensable *can't work*. Truth must be "across the board," or it fails to be *truth* at all.

the universe such a law exclusively holds of. Additivity again tells us that the application of such a law indicates the effects *due to gravity*: it doesn't require there to be no other effects.

25 Consider ordinary ways of speaking: "*A* is true, but I might be wrong about *A*," is incoherent. I *can* say that I'm open to changing my views about the truth of *A*; I can say that although *A* is true, in the future I may not believe *A*. (Admittedly, this sounds like I'm allowing myself to be wrong in the future – but how else am I to hang on to the use of my truth-predicate with respect to an *A* that's indispensable to my current beliefs, *and* express my openness to changing my views *at the same time*?)

26 See Horwich (1990).

This means that the minimalist view may fail for want of contrast: either the minimalist must design a metaphysically robust notion of "truth" not constructed from the minimal truth-idiom – not at all an easy task – or he must gesture wordlessly towards a metaphysically rich unexpressive notion of truth. I'm banking that neither move has much to recommend it.[27]

The second loose end concerns scientific talk of explanation. Both van Fraassen and Cartwright employ a notion of explanation that is not factive: explanations do not need to be true. But I think that this is a highly artificial notion which does not correspond to how it is used by scientists (or, for that matter, by ordinary people); philosophers find it more natural because of a long familiarity with nomological-deductive notions of explanation (which treat explanation along the model of validity rather than that of soundness), and because of Harman's inference to the best explanation – which seems to suggest that a number of explanations compete for first place (where the reward of being in first place is "being true").

Van Fraassen (1980: 98–100) offers a number of ordinary examples where a theory is treated as explanatory despite full knowledge that it is false: "Newton's theory explained the tides, Huygen's theory explained the diffraction of light ....".

None of these considerations should be seen as conclusive, because they overlook how explanation operates pragmatically. If you ask for an explanation of A, and I tell you something, B, and then add that B is false, you can respond: "B *would* explain A, if B were true." If I have been using B as an explanation of A, and subsequently learn it is false, I stop doing so: I can no longer offer it to others *as an explanation*. This doesn't prevent us from evaluating how well different theories *would do* as explanations if they were true (call this an evaluation of their "explanatory power"), and no doubt ordinary language can be sloppy about whether we're taking Newton's false theory to have explained the tides or to be something that *would have* explained the tides had it been true. Similarly, when engaged in inference to the best explanation, one can compare theories in terms of their explanatory power, and then decide that one is true (and therefore *provides an explanation of the phenomena*). Use of the subjunctive should not be confused with use of the indicative, even though ordinary usage can slur over such matters. Explanation, thus, *is* factive; explanatory power *is not*.

My purposes, in this and the last section, have been somewhat critical. The problem, in my view, is that philosophers have been gripped by the wrong issues. The central insight motivating so much of the work cited in

---

27 Still, does a commitment to the *truth* of fundamental scientific laws have to be a commitment to the hierarchy of natural kinds that's involved in science: the molecules themselves, for example, that Maxwell is concerned with? I probe this concern in Part IV.

    I have more to say about truth and various kinds of minimalism: see Part II § 7.

the last two sections is this: laws of a science are not brought to bear on the world or on the laws of other sciences by means of anything like the derivations we find in mathematics. I have called this the failure of theoretical deductivism.

We find a truncated version of this insight in use among those who focus on what is called the "many–many problem" or the "multiple realization problem." But narrow attempts to salvage physicalism by notions such as supervenience, and related concerns with property reduction, obscure the fact that the many–many problem is a specific instance of something far broader.

If philosophers had recognized how broad the failure of theoretical deductivism is, and how widespread is the use of idealizational techniques for circumventing it, they might have found their way to a "physicalism" that simultaneously honors the autonomy of the special sciences: In short, they would have seen exactly in what sense actual physicists are physicalists, and why *this* physicalism is compatible with the continued existence of specialized vocabularies in the specialized sciences. Making this clear was the motivation of § 4.[28]

A tendency to ply epistemological insights into metaphysical truth *may* explain why philosophers such as van Fraassen and Cartwright deny the truth of scientific law for reasons having to do with considerations of confirmation or observation. A quick meditation on the role of the truth-predicate was therefore called for to deflect this strategy.

My discussion of Cartwright, thus, illustrates that the idealized bridges needed between fundamental laws and the world do not imply such laws are false. It is also possible to assume that the idealized bridges themselves indicate what the world is like, or to assume that when scientists use such idealized bridges, they are taking them literally. I have not criticized this move explicitly on the part of any particular philosopher, but it is possible, and does occur in the literature.

Up to now, I have been concerned primarily to remove various philosophical obstacles to a clear view of gross regularities. I now offer a positive exposition of my own views about these regularities and their importance to scientific practice.

---

28 Notice, from a metaphysical point of view, how little the failure of theoretical deductivism changes: truth is untouched, explanation seems largely untouched, e.g. the status of the D-N model of explanation is unchanged by the failure of theoretical deductivism (a result that, *a priori*, may be surprising, given how important deduction is to that model), metaphysical issues about whether or not the natural kinds spoken of in special sciences are physical, are also unaffected by the failure of theoretical deductivism. The general lesson is this: whether theoretical deductivism is true or not has an impact primarily on the epistemology of the empirical sciences, and not on their metaphysics.

§6

GROSS REGULARITIES

In § 3, I described the gap between scientific theories and their domain of application as a *boundary layer*. One reason this layer is overlooked is that it is clear that science pretty regularly gets successfully tested and applied. Consequently, whatever is involved in the boundary layer is often successfully navigated by those who want to test their theories or apply them. Let's examine the tools available for doing this.

Consider the middle-sized objects we interact with on a fairly regular basis, objects we (naively) regard as commonsensical or observational: stuffs (containers of milk, lumps of gold, vials of blood, etc.), various living beings (people, squirrels, poinsettias, etc.), mechanisms (electron microscopes, batteries, bridges, etc.), and so on.

The irreducible status of (many of) the laws of the special sciences to physics, and of (many of) the laws of those special sciences to other special sciences that (metaphysically speaking) underlie them, has its analogue in the irreducible status of a body of knowledge about middle-sized objects that, strictly speaking, does not belong to any science; we, as lay-persons and as specialists, rely on a body of knowledge about such objects that is only partially reducible to *any* science.

That this is true of a great deal of what we know about people and animals is widely known. But it holds, too, of the inorganic world: This is why engineering is not merely theory-driven by the principles of physics, chemistry, and other special sciences, but requires hands-on experience and craft.

What hands-on experience and craft teach are general truths, *gross* regularities. Some of these regularities are expressible in language, but many are items of kinesthetic "know-how." I'll now give a few details about each kind of gross regularity.

Articulated gross regularities

These regularities, which are expressible in language, hold more or less much of the time under certain circumstances which can be only partially specified. They cannot, as they stand, be hedged into laws because, usually, they have small domains of application, and admit of too many exceptions which can be recognized only during the process of empirical application. Furthermore, they often cannot be derived from laws already in place in one or another science because of sheer scope problems: there are lots of orphaned gross regularities we recognize as appropriate to particular sciences, programmatically speaking, although tractability prevents their derivation.

Here's an example (Davidson 1963: 16). It's a good bet that a baseball thrown at a window will break the window. If this regularity were to be derived from the principles of a science, no doubt that science would be physics. But due to the nearly infinite physical variation possible in windows, baseballs, and velocities, the needed derivation is out of reach.[1]

Many articulated gross regularities, though, inhabit a programmatic no-man's land: they are not matters *merely* of chemistry alone or of physics alone, but are properly located in the domains of several sciences at once – without, however, it being possible to derive them from the joint-application of the principles of those sciences.

Here's another example. Most of us are satisfied with the rough-and-ready description of how to use *Elmer's Glue*, found on the back of the bottle. But working with even a small number of materials (vinyl tile, plastic, wood, glass, paper, etc.), under varied circumstances, one finds that, although general chemical and/or physical properties governing kinds of glue are rather easy to sketch, this rarely tells us how specific applications will go; that is, how factors (eccentricities in the materials, atmospheric effects, differences in the specific make-up of the batch of glue being used, and so on) affect the life of the bond and its capacity under different stresses.

We can't predict glue's behavior to this requisite degree of delicacy because the physical and chemical details are too complex: our primary avenues to knowledge[2] here are hands-on practice with the glue in a wide variety of situations, or the wise guidance of an experienced friend. Furthermore, the large body of gross regularities learned this way, regularities about the behavior of a particular glue in respect to particular materials in particular circumstances, are not regularities that need hold under circumstances closely resembling the ones in which we have learned they do hold: Small perturbations in application conditions can cause unexpected effects. For example, small differences in weather may affect the bond; or we

---

1   For the same reason, definitions of "baseball" and "window," in physicalistic terms, are out of reach.
2   I say "primary" rather than "only" because, often, scientific results can be brought to bear on something to successfully predict its properties in certain circumstances.

may find the glue works for a kind of wood, although it doesn't generally work with wood, or that it works better with a kind of wood if we dilute the glue with a particular solvent; and so on. Gross regularities about glue cannot be expected to obey nice thresholds of application – where these are taken to be either chemical parameters or physical ones.

Is *glue* special? Hardly. The research that generates useful products such as glue is more *laboratory-driven* than it is *theory-driven*, for it often must proceed by guesswork, hunches, and luck. This means that it usually proceeds by the (often accidental) discovery and exploitation of gross regularities. I mean this both in the sense that we commonly rely on gross regularities we discover in the lab situation to develop the product in question and in the sense that gross regularities we discover when testing the product empirically are pertinent to utilizing it later (as I just illustrated with *Elmer's Glue*). Often the gross regularities that are discovered (and make a certain product possible) may remain gross regularities for the indefinite future: we may not subsequently come to understand quite why, scientifically speaking, the gross regularities in question hold.

Accidental discoveries of gross regularities not predicted on the basis of theory litter the annals of invention. Crucial aspects of the vulcanization process were discovered by accident. Another example is the accidental discovery, in 1922, by T.C. Midgley Jr and T.A. Boyd that adding small amounts of tetraethyllead to gasoline reduces "knocking." *How* it does this, the *mechanism* of the reaction, is *still* not definitively understood. Other examples (arising during the development of the steam engine and metallic conduction) may be found in Hacking (1983: 162–5).[3]

Note the following:

1   Even when scientific theory dictates the possibility of a certain result or product, the success of the application still largely depends on details found only by direct testing in the application situation (e.g. drug testing).

2   These facts, about the intrinsically empirical character of applied science and the consequent need to develop and utilize gross regularities, hold of the applications *of any science*, not just of chemistry and physics. The evidence for this is that just about any designed product needs considerable testing (and often by customers!); no collection of scientific theories in hand makes a product *a priori* trouble-free (i.e.

---

3   A feel for this absolutely ubiquitous phenomenon can be acquired from a cursory reading of certain articles that appear regularly in *Scientific American*; for example, Neher and Sakmann (1992); Snyder and Bredt (1992); and Thomas (1992). Interestingly, one also gets a feel for it from investment magazines such as *Financial World*, because investors, naturally, are interested in getting a sense of the developmental situation for certain products. See Kindel (1992), for example.

totally *predictable*). One can't even predict how (most) small changes in the shape of the handle of a screwdriver will affect how user-friendly it is, despite our utter familiarity with both hands and screwdrivers.[4]

Notice what I am *not* saying: I am not saying that *every* articulated gross regularity resists full incorporation by some science (or other). That clearly isn't true, since problem-solving sciences routinely extend their scope by, among other things, incorporating gross regularities, suitably modified, into their corpus – providing *explanations* for them, as it is put. My claim is only that a large number of such gross regularities successfully resist such incorporation, will continue to, and that, despite this, we need to rely on such regularities incessantly when we apply science. This explains in part why the *application* of science (any science) to solve problems posed in our daily lives is not a mere corollary of pure scientific doctrine, but a full-time job in itself.[5]

Unarticulated gross regularities[6]

In addition to articulated gross regularities, there is a gigantic class of gross regularities, knowledge of which is never articulated linguistically. Among these unarticulated regularities are the techniques acquired when one learns

---

4   This example points towards our need to utilize unarticulated gross regularities, of which more shortly.
5   For a nice illustration of the theme of this last sentence, see Basalla's discussion (1988: 92–101) of the development of the atmospheric steam engine and radio communications.
     Suppe (1989a: 282– 3) writes:

> What engineering tries to do ... is first determine a worst-case ... circumstance the device might have to operate in, where this worst case is sufficiently simple that auxiliary hypotheses can be developed which will predict how a particular design will fare in it. One then theoretically overdesigns one's prototype to work reliably in this worst-case circumstance and tests the prototype with relatively low risk of failure. But such overdesigns are expensive and inefficient, so one then experimentally tries a cumulative series of design modifications that gradually eliminates the "overdesign," subjecting the modifications to testing in a variety of normal-use situations, until one comes up with a sufficiently modified design that is both reliable and economical. Such "trial and error" design improvement is at the heart of engineering research and development, and it is guided more by the intuitions and experience of designers than by accurate predictions rooted in established scientific theory.

     The substance of, as it were, "intuitions and experience," I take to be knowledge of gross regularities.
         Also, see Smith and Mindell (forthcoming) for an extended case study of the development of the turbofan engine, which illustrates the themes of this section.
6   A discussion somewhat compatible with my take on unarticulated gross regularities may be found in Polanyi (1958: Chapters 4–5).

crafts like carpentry or car-mechanics, what is called "how-to" knowledge or "demonstration" knowledge. For example, a carpenter, showing you how to plane a piece of wood, might say: "Good God, no! You've got to hold it *like this!*" (And that might be as much articulation as the indicated gross regularity can get.) A car-mechanic may "explain" how he started your car this way: "I always jiggle the whosididst *like this.*"[7]

This kind of knowledge, what we may call a *feel* for the interactional patterns of things, extends beyond our feel for our tools and the objects around us: it also includes a (nearly subliminal) feel for what people or animals we know are likely to do. Professionals such as con artists or policemen have a sense of how (most) people will act in certain circumstances (at least, they do if they're any good).[8]

A subliminal reservoir of unarticulated gross regularities is precisely the kind of knowledge (skill) engineers and experimenters cultivate: They learn to "jiggle," or otherwise modify in small ways, the devices and objects they work with to steadily improve an experimental result or product. This kind of knowledge is as problematical and limited in scope (each new device brings with it the need to learn some of its quirks) as is knowledge of articulated gross regularities, but the question of how such regularities are to be incorporated into pristine empirical science doesn't arise, because such knowledge is, pretty much, not described but only demonstrated.

That is not to say such unarticulated gross regularities can *never* be articulated. One way of doing so is to introduce new vocabulary. New words are coined not only for parts of devices but even for certain regular ways such devices malfunction. Nevertheless, many such gross regularities remain unarticulated because they are too localized to bother with, or because our grasp of them is too kinesthetic to be translated into a linguistic form.[9]

---

7   Another nice example is how we learn, almost subliminally, to manipulate a key to open a lock. We forget how much learning is involved, and that it is largely incommunicable, until circumstances force someone else to open the door for the first time, and she complains that "the key doesn't work."

8   See Goffman (1971, in particular, p. 241 note 2); but illustrations crop up throughout his essay.

9   Two delightful examples of the impact of unarticulated gross regularities on technology may be found in Basalla (1988). The first (p. 83) goes this way: Americans in the late 1700s wanted a textile industry independent of the British industry. But they were unable to design the up-to-date machinery the British had, for

> [w]hat they lacked were designers of textile machinery and workmen who knew how to adjust, control, and maintain those machines so that yarn and cloth of an acceptable quality could be produced in quantity from native wool and cotton. Having the actual machines on hand did not suffice if there was no one experienced in assembling and using them. This the Americans learned in 1783 after several key textile machines were smuggled into Philadelphia from England in a disassembled state. After four frustrating years, during which no one competent could be found to assemble them, they were shipped back to England.

Sometimes, unarticulated gross regularities can be modeled scientifically despite being inarticulatable in ordinary language. Consider the skills needed to ride a bicycle. A simplified physical model can be constructed of the forces acting on a bicycle, and of the physical adjustments a rider must make to stay balanced; and computer graphics can visually depict the evolution of such a model over time. It won't capture everything (bicycles have individual variations to which riders adjust, and even mediocre riders like myself can navigate over difficult terrain in ways that go beyond what we can success-fully model), but this illustrates how visceral know-how can sometimes be incorporated into science, bypassing a verbal detour in less formal terms.[10]

Five remarks about articulated and unarticulated gross regularities

(1) Some might think that articulated gross regularities are no different from the laws of the special sciences, since (a) the *ceteris paribus* clauses of such laws mark out a more or less *open* class of exceptions, and (b) the language of such laws is often as qualitative and vague as is the language of genuine gross regularities. It is perfectly alright, I think, for the distinction between articulated gross regularities and the laws of (certain) special sciences to be somewhat fuzzy, especially since the epis-temic phenomenon that gives rise both to gross regularities and to the semi-autonomous laws of the special sciences is the same. But there are important differences. For a subject matter to be sensibly regarded as

He concludes: "Successful transfer of textile technology was not achieved until experienced British emigrant artisans were able to put their nonverbal knowledge to use and produce the machines for the American manufacturers."

The second example involves British attempts to steal the secrets of Italian water-powered throwing machines used by the Italian silk industry. Despite the availability of detailed pictorial depictions of the device, success came only when the "industrial spy" John Lombe visited Italy and familiarized himself with the machine over the course of two years (see Basalla 1988: 84–6).

Other examples may be found in Polanyi (1958). One such concerns the scientific study of spinning cotton (to catch up on "what the spinner knows"). Polanyi adds: "I have myself watched in Hungary a new, imported machine for blowing electric lamp bulbs, the exact coun-terpart of which was operating successfully in Germany, failing for a whole year to produce a single flawless bulb" (p. 52). There are, of course, numerous contemporary examples.

10 I have described the unarticulatedness of such gross regularities in terms of whether there is a *public* language in which to express them. My concern takes this form because it is the absence of such a public language which requires the communication (when possible) of unarticulated gross regularities to occur by means of gesture and demonstration. One may wonder whether the able practitioner's mastery of unarticulated gross regularities is *repre-sented* psychologically. This is a question on which I have nothing to say at present. I believe the answer to it does not affect any of the issues before us.

the domain of a science, it should be at least partially theory-driven (by the principles of that science); otherwise it is just a collection of "interesting facts." The special science should have a program and scope: Some of the claims in the science should guide predictions as well as provide deep-seated explanations for a certain class of phenomena. As examples, consider the laws and models utilized in economics and evolutionary biology,[11] or the periodic table in chemistry. Gross regularities, by contrast, are just a loose collection of general facts that have in common only our need to rely on them when applying and testing scientific doctrine, or their capacity to arouse our curiosity about what is causing them.

(2) Included among gross regularities are the "commonsense regularities" G. E. Moore pointed to as examples of knowledge: "Pencil marks on paper stay where we put them"; "People like to eat and drink fairly regularly"; and so on. These regularities have long been recognized as peculiar. They never seem to be tested, and are somehow epistemically foundational in nature.[12] But, as I have indicated, gross regularities are actually a broader class of regularities which includes much more that only specialists know: that certain alloys can handle only so much tension; that certain dyes cannot be used on certain fabrics (without untoward results); that certain animals prefer certain habitats; that "Calcium often acts by binding to a ubiquitous cofactor called calmodulin."[13] Such regularities are not *common* knowledge at all.

(3) Although my introduction to the topic of gross regularities, and some of the examples I have chosen, might give a different impression, nothing in the definition of "gross regularity" *requires* a restriction to objects we pretheoretically regard as commonsensical or observational. Gross regularities include unexplained regularities with regard to items such as electrons or black holes.[14]

(4) What I mean by "gross regularity" is rather close to what Hacking (1983: Chapter 13) means by "phenomena." The terms are not identical, though, because I'm concerned with relatively repeatable events. Experiments, in Hacking's sense, are examples of phenomena. But he argues (correctly, I think) that because experiments are continually refined, they are for the most part not repeated. Experiments, even if we took them to be repeatable events, are too large to be gross regularities in my sense: Rather, they are *made up of* gross regularities. It is by getting a grip on

---

11 Kauffman (1993) offers a program powerful enough to distinguish between biological correlations which are due to selection, and correlations present *despite* selection.

12 Commonsense regularities also underpin evidential practices in mathematics. See Azzouni (1994: Part 3 § 2).

13 Snyder and Bredt (1992: 71).

14 I'll take a closer look at these sorts of gross regularity in Part II § 5.

the (mostly unarticulated) gross regularities about materials and apparatus in an experimental situation, and forcing new ones to emerge, that an experimenter makes the requisite incremental improvements in an experimental situation. Gross regularities are, as it were, among the pieces we tinker with and modify, and recognize new examples of, as we improve experiments. Without such pieces, it would be impossible to make any improvements at all to experiments (a successful experiment won't spring full-grown from the lab the way Athena sprang out from Zeus' head).

(5) Gross regularities, especially the unarticulated ones, have been acknowledged implicitly under the description "tacit knowledge," for instance, by Basalla (1988), Collins (1992), and the already-cited Polanyi (1958). But tacit knowledge in this sense is often treated as an unanalyzable social skill that a scientist or craftsperson acquires, and some contrast it explicitly with the "algorithmic knowledge" which successfully applied science gives us. Furthermore, this apparent "black-box" quality of tacit knowledge allows a charge of capriciousness.[15] By contrast, describing tacit knowledge as simply the acquiring of unarticulated and articulated gross regularities, allows us to see why it must be transmitted from practitioner to practitioner by demonstrations, why failures of transfer take place, and why it is limited in scope the way it is. Also, I avoid the implication that tacit knowledge is merely cultural in nature – arbitrary beliefs one acquires by belonging, as it were, to a guild – as opposed to being objective knowledge about the world and its workings.[16]

I can now answer the question this section opened with: how, in the application and testing of empirical scientific theories, do we cross the boundary layer separating scientific doctrine from the domain it is supposed to apply to? We brave its turbulence, armed with gross regularities.

---

15 See e.g. Collins' *Proposition One* (1985: 73). He distinguishes a view of knowledge as "the sort of information that enables a computer to carry out its programmer's intentions" – what he calls algorithmic knowledge – from knowledge "as being like, or at least based on, a set of social skills" (p. 57). I resist assimilating knowledge of articulated and unarticulated gross regularities to tacit knowledge in this sense. Knowledge of gross regularities fails to be algorithmic in a technical sense only because it might be inexpressible in the language we use. But this doesn't mean knowledge of gross regularities is not applied mechanically precisely as one applies what Collins would regard as algorithmic knowledge.

By the way, experimental skill is often kinesthetic the same way athletic abilities are. This goes a long way towards explaining difficulties of transmitting experimental skills from one person to another. Imagine a professional pitcher attempting to show me how to pitch the way he does. The result fails because my seeing how he moves his arm, or thinking I see this, will not enable me to move my own arm similarly. Thus, just because something cannot be taught does not mean it is not knowledge – and knowledge about the world in as strong and objective a sense as anything else we are willing to call knowledge.

16 That is, they offer "thick epistemic access" in the sense explained in Azzouni (1997a).

§ 7

PROCEDURES AND
PERCEPTUAL PROCEDURES

That gross regularities are epistemically independent of the sciences (that gross regularities live within the boundary layer between science and the world) bears directly on the nature of our evidential practices and, more generally, on the methods by which we pick out what our terms refer to. I now show how.

Consider gold. Early techniques for recognizing the stuff were simple and utilized no special instruments. To verify something was gold, one bit into it, weighed lumps of it in one's hand, checked its color, or submerged it in water (if one was especially clever). This has changed. Some tests for gold are quite refined and highly specialized, involving chemicals and/or devices of various sorts. The result is that we find gold in so many more places nowadays than people used to. We have tools and instruments for recognizing its presence in chemical compounds and distant celestial objects. Even the minute quantities of it in seawater can be detected.

I call *procedures* the methods by which we identify the things our terms refer to. Such methods exploit and forge causal connections between ourselves and the objects they are used to recognize. As my talk of "tools" and "instruments" is supposed to suggest, we should not understand "procedure" too crudely. If we consider ordinary middle-sized objects like tomatoes, we naturally think of a small number of hands-on ways of identifying them, such as tasting them, cutting them open, and so on. But with objects outside our sensory range, things are more complicated. A single procedure may apply several different devices, as well as mathematical calculation (the procedures for recognizing subatomic particles or synthetic elements often look like this), not to mention various chemical processes.

Generally speaking, a group of procedures used to causally connect us to a kind of thing is a heterogeneous collection: They are sequences of *ways* of applying devices, certain chemical processes, etc.[1] The devices involved may

---

1   Here is an example from Haj-Hussein *et al.*: "Westerlund-Helmerson ... analyzed zinc and lead oxides for chloride by boiling the sample with silver nitrate and nitric acid and

be able to detect the presence of any number of quite different kinds (as, for example, a spectrograph can). In order to understand *why* the procedures are associated the way they are, we need to know what theories users have about the kinds they are trying to interact with. But it is a matter of brute fact, independent of the beliefs held by the users of these procedures, whether or not that collection of procedures does indeed (most of the time) interact with a scientific kind.

The reason that, in general, the same substance can have so many hetero-geneous procedures associated with it is that substances, generally, do not occur in one context where their properties make it easy to pick them out. They occur in minute quantities *all over the place*, combined in various ways with all sorts of things, and it usually takes intricate and sophisticated methods to determine their presence. A glance at some chemistry journals shows that one ongoing professional activity of chemists is the development and refinement of procedures for determining the presence of various substances in special circumstances.[2]

Despite the almost unmanageable complexity in types of procedure available, there does seem to be a simple distinction between procedures involving a hands-on operation unaided by any form of instrumentation – *perceptual procedures* – and the rest, which are *instrument-enhanced procedures*.[3]

This distinction is epistemically important; and I'll show this in the rest of this section. Then, in § 8, I'll turn to the (practical) possibility of eliminating the distinction, and what, if any, implications for epistemology such an elimination would have.

Here's why the distinction is important: given our *current* biological and technological makeup, perceptual procedures are peculiarly *central* to our

---

dissolving the washed chloride precipitate in ammonia. The silver in the dissolved precipitate was then determined by atomic absorption spectrometry" (1986: 38).

    A standard textbook of chemical instrumentation (Strobel and Heineman 1989: 9) explicitly adopts a modular approach to instruments: any instrument has two sequences of modules, first, a set of characteristic modules (these, properly speaking, gather data) and a set of processing modules; the latter can include amplifiers, microcomputers, and so on.

2   Here is a small sampling from only two issues of two journals, and the interested reader can easily find hundreds more: Eiceman *et al.* (1986); Haj-Hussein *et al.* (1986); Rapsomanikis *et al.* (1986); Rouchaud and Fedoroff (1986); Sánchez and Blanco (1986); Stoytcheva (1992); Wirsz and Blades (1986); Xingguo *et al.* (1992). Notice that in many cases, the title says it all.

    Chemistry is the best place to get a feel for how complicated it is, in general, to recognize that some of *A* is located at (in) *B*; that is, how complicated the causal relations are that must be forged or exploited to connect us to those quantities of *A* located at (in) *B*.

3   I sometimes speak of the application of perceptual procedures to test something as the application of "pre-scientific procedures," or "pre-scientific methods." *I* mean to indicate by this term both that such epistemic practices predate empirical science and that they are, epistemically speaking, prior to methods available in empirical science. For a justification of this terminology in my sense, see what follows.

epistemic practices. For, first, *any* procedure we use must contain perceptual procedures in order to be useful. That is, *whatever* the distal causal mechanisms are that a procedure exploits, when it gets to our end of the causal chain, there had better be flashing lights, beeps, verbal remarks, words on a page, computer printouts, or something else our senses can recognize.[4] Second, because of the small number and limited powers of our naked senses, the distinctions we can make using them are small in number (relative, of course, to what is out there). Perceptual procedures, therefore, are similarly limited and come up repeatedly in evidential contexts. Third, because perceptual procedures operate among middle-sized objects, complex items (ourselves included) to which it is extraordinarily difficult to apply the sciences in any sort of detail (as we've seen), such procedures are built out of gross regularities; that is, our use of them goes on more or less independently of the sciences.

I illustrate this briefly. Consider eyesight. We have learned how to gauge distances between objects, and between objects and ourselves, by means of what we see; and we have also learned to distinguish colors by the same means. Some of the capacity to make these distinctions is, admittedly, hardwired,[5] but not all of it is. Learning things of this sort is learning a body of *unarticulated* gross regularities which we instinctively apply when we look about. Like the application of any gross regularity, they can misfire, and a good example are untutored estimates of the distance of the sun from the earth.

That such perceptual procedures operate independently of the sciences seems relatively obvious, and I wouldn't spend much time on it except that many philosophers deny it. Consider, for example, Lakatos (1976, p. 107): "calling the reports of our human eye 'observational' only indicates that we 'rely' on some vague physiological theory of human vision."

"Physiological" is a stretch in any case, since many people (and entire cultures) have never even considered the possibility of a theory of human vision – of any sort – let alone relied on a vague one. The gross regularities we utilize about our senses are general (largely unarticulated) claims about when they can be trusted and when they can't: These regularities are used in our evidential procedures, and are needed to justify whatever physiological theory we've managed to eke out, not the reverse. (We don't have much of an understanding of why certain optical illusions work, but we still know when to discount them.)[6]

---

4   See Strobel and Heineman (1989: 9), where "display or readout" is included among the "processing modules."

5   See Hardin (1988), especially the discussion on object metamerism, adaptation, and contrast (pp. 45–52).

6   Philosophers are regularly prone to an overestimation of the reach of scientific theory, as we saw in § 3, and often the effect of this is the creation of scientific theories where none

Here's an illustration of the use of pre-scientific methods. Consider the act of writing on a piece of paper. It's true that such a process need not have been reliable. For it could have been that, fairly often, paper affects the marks on it so that supplementing our memory with a paper trail is risky (something leading to mistakes often enough to warrant not doing it). How have we ensured that this doesn't happen?

Well, we have used *perceptual procedures* to measure the reliability of the products (such as paper and pencils) we work with. Such methods were available long before the flourishing of empirical science, and they are methods, as we have seen, which are epistemically prior to the knowledge and tools of empirical science. Like most of us, I no longer use paper and pencil (very often). Instead, I word-process, and have developed confidence in floppy discs because what is recorded there has matched printed copy when I check it later (by eye). This confidence has nothing to do either with scientific doctrine (for I know virtually nothing about how floppies work), nor with the authority of experts (for I don't rely on any experts in this area to justify my confidence on these products). It is simply that I have checked the reliability of floppy discs in a pre-scientific manner – that is, I have checked *by eye* the results against other methods I have for recording data – and I am convinced they are reliable.

That these methods (of checking epistemic claims about objects and processes we have direct access to) are, as a group – independent of, and prior to, justification in the sciences, can be recognized by observing that the idea that scientific developments could somehow show that all such methods are systematically misleading is virtually incoherent. For *were* all such methods misleading, we would have no way of testing scientific claims either: All such claims are tested, *and can only be tested*, by processes with indicators pre-scientifically recognized by perceptual procedures. If positrons have certain properties, that is evidence for or against certain theories. But such evidence is useless until translated into a form we have access to: certain lights must flash. And, in turn, to be sure that we've seen what we

exist. Resnik (1989: 133), when discussing ordinary (pen-in-hand) calculation, writes: "computational inference ... presuppose[s] some theory about ourselves and our mathematical training." Of course, there is no such *theory*. He then goes on to discuss electronic calculators as if they are related epistemic phenomena. This, alas, is simply not true. Electronic calculators *are* accompanied by scientific theory; human calculators are not. In both Lakatos (1976) and Resnik (1989) there is a tendency to blur the epistemic differences between a dependence on empirical *science* and a dependence on empirical *pre-science*. Of course, they are hardly alone. Guilty of the same conflation are Churchland and Quine. The mistake, I surmise, is due to an uncritical acceptance of theoretical deductivism.

Deductive holists might again claim that I construe "theory" too narrowly. Recall my response in the Introduction.

think we've seen (a red light flashing several times, and not a green light flashing once), we rely on perceptual procedures.[7]

Were the results of our pre-scientific testing of floppies against other methods of recording data to yield the result that floppy discs are entirely unreliable in contradiction to the results of science, *the science would yield*. Even if no scientific reason were found for why such things are untrustworthy, no one would use them. My point is that this would not be irrational. On the contrary.

Let's pause a moment to summarize and make clear what the arguments here are relying on by way of assumptions. First, there is *the undercutting argument*. Perceptual procedures are evidentially central *biologically*,[8] so scientific development, however it goes, cannot undercut their veridicality in evidential contexts without infirming the very methods that justify scientific claims themselves. This argument *does not* rely on perceptual procedures utilizing gross regularities, nor does it rely on the relative independence of gross regularities from the sciences; for even if scientific knowledge *could be* brought to bear on perceptual procedures and the gross regularities associated with them in an intimate way, this knowledge would still have to vindicate the epistemic veridicality of the application of gross regularities and perceptual procedures to evidential contexts for fear of overthrowing its own evidential basis. Of course, a sceptical collapse of the entire structure, science *and* its evidential methods, is still allowed by these considerations.[9]

However, that gross regularities and perceptual procedures are (relatively speaking) independent of the sciences *is* used to explain why, epistemically, perceptual procedures operate in their own domain as it were: testing them, and recognizing their range of application, goes on largely independently of science.

I close out this section with a few last observations about perceptual procedures. Notice, first, that nothing like incorrigibility holds of perceptual procedures. Even if science plays no genuine role in evaluating the epistemic value of our sight in evidence gathering, it does not follow that we cannot make mistakes using our eyes. Such mistakes, though, are corrected, not by the top–down application of scientific theory (or, at least, not very

---

7   This is a simplification. In practice, the scientist does not always have something as simple as flashing lights to look at when evaluating data. For example, she may have to stare at streaks in photographs, and it may take years to learn how to read information out of such things.

8   Well, *now* they are, but surely, with sophisticated genetic engineering almost upon us, perhaps things could change. I address this in the next section.

9   Notice the qualification: the undercutting argument does not justify perceptual procedures, or gross regularities used for such, *tout court*; it justifies such things only *as applied in the evidential context*. Because of this, as I show in Part II, what is evidentially required are not sentences or truths as traditionally construed.

often), but rather by the further refinement of the perceptual procedures themselves. If I think I've made a mistake about what I'm seeing, I move in for a closer look;[10] and doing so helps me evaluate first impressions. Even though perceptual procedures provide an evidential foundation for scientific practice, they are not an epistemic foundation in the sense, say, that sense-data, if such existed, would be. Rather, the picture is closer to the sort of picture sketched by epistemic coherentists, except that the coherent whole is not all our knowledge but rather just the group of gross regularities and perceptual procedures relying on them that are used in evidence gathering.

Second, it certainly has not been shown that perceptual procedures are *entirely* independent of the sciences. This couldn't be right, because gross regularities themselves are not entirely independent of the sciences; it can certainly happen, now and again, that scientific doctrine corrects results based on perceptual procedures. But this weak concession is hardly one the deductive holist can use.

Third (and I illustrate this in detail later), the view taken here does not allow a definition of "observational terms" so that those terms are semantically independent of scientific theory. I'll say this much now: that perceptual procedures are epistemically special need not be implicitly acknowledged in ordinary usage, nor does this special status depend on their being so acknowledged. The common terms which are the best candidates for being terms which refer to perceptual procedures are terms such as "see," "touch," and so on. But these actually do not serve this role, since ordinary usage clearly allows collateral information to influence what someone can "see," and collateral information can include the results of instrument-enhanced procedures, e.g. someone looking through a telescope and "seeing" a mountain on the moon.

Those in desperate search of observational terms in ordinary language have tried to argue that, in cases like this, one is not "actually seeing" the mountain. This lands them in the undesirable position of sorting collateral information on the part of viewers from what is "actually seen" (after all, collateral information involves inference, not *sight*, or so it seems). I don't need this distinction: perceptual procedures can be present and be epistemically significant even if *no* terms of ordinary language unequivocally describe their operations.

Last point. The distinction between perceptual procedures and instrument-enhanced procedures turns on whether in fact the instruments involved are being used *as* instruments, and thus on the ontological commitments we have about what out there can be reached by means of these instruments. The distinction, that is, is a distinction drawn in naturalized epistemology (within our own conceptual scheme), and so is sensitive to *what* we take the

---

10 I'm myopic. Others might step back.

individual to be examining. If someone looks at microbes through a microscope, he does so by means of an instrument. But if someone peers into it to see the "pretty colors," or "blurry shapes," she is using only perceptual procedures, and not an instrument-enhanced procedure, despite looking through a microscope. Since any instrument-enhanced procedure needs perceptual procedures at the user-end, it is always possible for the user to "see" the results of the perceptual procedures rather than things more distally available.

We do sometimes describe an ignorant person peering through a microscope as someone who is "really seeing microbes, even though he doesn't know it." What makes this talk appealing are our own ontological commitments, here, our commitment to microbes as opposed to the relatively vague "blurry shapes," or "pretty colors." We often cut causal chains between us and objects we perceive at the "ontological joints," even if to do so is not psychologically accurate.[11]

---

11 See Part II § 3 for further remarks about the ontological status of "objects" of perceptual procedures.

§ 8

SHEDDING PERCEPTUAL
PROCEDURES

There are two objections to the distinction between perceptual procedures and instrument-enhanced procedures that I should respond to. The first is that it looks like I'm pushing a sharp distinction between perceptual procedures and instrument-enhanced procedures. But certain examples seem to show no such (sharp) distinction is available. The second is that I have given sense-perception a central role in evidence gathering, and stressed the necessity of using these senses when gathering evidence, despite the possibility that we could develop new senses (or modify our old ones in such a way that their scope and limits change drastically). I take these worries in turn.

It may seem obvious we cannot so simply divide instrument-enhanced procedures from perceptual procedures. Many of us wear eye-glasses or contact lenses nearly all the time. Does this mean that nearly *every* procedure we apply (with our eyes open, anyway) is instrument-enhanced?

A definition will help. The use of an instrument is *epistemically conservative* if all the distinctions it provides are pre-scientific ones. Binoculars are epistemically conservative: they enable people to see things at a distance others can see with the unaided eye if they are closer. A microscope is *not* epistemically conservative: a mere change of position will not enable anyone to see microbes with the unaided eye. I modify my initial distinction between instrument-enhanced procedures and perceptual procedures by describing a procedure as instrument-enhanced only if the instruments involved are *not* epistemically conservative.[1]

Now I can respond to the worry raised two paragraphs back. As far as individual knowers are concerned, and as my distinction was originally presented, yes, the use of eye-glasses can certainly mean that *every* procedure implemented is instrument-enhanced (although I will raise a caveat in a moment). But this isn't true of our evidential practices *as a group*. Eye-glasses as far as our epistemic practices *en masse* are concerned are epistemically

---

1   This introduces a sociological dimension: whether something is epistemically conservative turns on what perceptual procedures are generally available to *other people*.

conservative: we eye-glass wearers use eye-glasses to make distinctions others make without them (and that even we often make by squinting). Eye-glasses correct a visual weakness, and consequently, epistemically speaking, they supply no more advantages to their users than the individual with naturally occurring 20/20 vision has anyway.[2]

Why is this point worth making in an *epistemological* context? Because what justifies the evidence eye-glass wearers gain through their senses is *not* a scientific theory such as optics, not the theory that explains how eye-glasses work, but precisely the *conservativeness* of their evidence-gathering powers, the fact that we can test their veridicality using perceptual procedures. An indication of this is that such items were in use for purposes of magnification long before a sensible optical theory was available. That glasses did the job one hoped they did was verified either by testing them against the visual experiences of non-users or by testing them against the perspective of the same user from a closer perspective.

Let's modify the case. Imagine a (near?) future where biotechnical sense-organs are available. Imagine that easily installed eyeballs can be purchased which provide enhanced sense perceptions. The cheaper ones allow "binocular" or "microscopic" vision. More expensive items allow the sighting of ultra-violet light, (large) microbes, or the black holes in the night sky. Won't the (modified) distinction between perceptual procedures and instrument-enhanced procedures collapse now?

Well, yes. But it won't change anything, epistemically speaking. Let me take a little time to show this by means of an analogy with the microscope. Hacking (1981) discusses the evolution of the microscope at fair length, and one important conclusion to draw from his discussion should be no surprise to the reader who has gotten this far. Microscopes, like other technical devices, were not developed solely in a top–down manner using scientific results. Rather, they were developed and implemented in the standard empirical fashion that necessarily involves the application of gross regularities. That is, their development involved all the kinds of sloppiness and hands-on manipulation that development is known for.[3]

But there are two other important points to make about the use of the microscope. The first is that the successful user does not need to know

---

2 Well, this isn't entirely true, as it turns out. My glasses, for example, give me 20/15 vision. But my use of glasses is epistemically conservative despite this, for I have checked their veridicality by using them to see a somewhat distant object, removing them, and then moving closer. There is still the possibility of instrument-enhanced senses which are not epistemically conservative. I address this next.

3 In fact, early developers may have gotten their hands on gross regularities that later utilizers of the microscope didn't have. Hacking points out that we don't know how, given the equipment at his disposal, Leeuwenhoek managed to make "such marvellously accurate drawings of his specimens" (1983: 192).

particularly much (if anything) in the way of physics to use it. The second is that, regardless of what she knows by way of theory – understood narrowly as physics, or broadly in terms of general knowledge about microscopes – the successful user must *still learn* to use the microscope by a hands-on process of interacting with what she sees through it. The user, that is, must learn to produce stable images, and to distinguish what is "really there" from what is an artefact of the microscopic process itself. This is not trivial, for it involves learning an indeterminate number of unarticulated gross regularities which tell the user when what is seen should be taken seriously.[4] The process involves acquiring knowledge independently of empirical science (in particular, of physics) despite being in the context of the use of an instrument the operation of which *is* (pretty much) explained by physics.[5]

Let us return to biotechnical eyeballs. As with the microscope, the utilization of such augmented senses will take on an epistemic life of its own. Although scientific doctrine will apply to how such senses operate in a broad way (pretty much as it applies to our understanding of our current sense-organs), users will largely be on their own as they fine-tune their abilities with their new senses and maneuver about by means of them. They will learn gross regularities that contribute to a nearly instinctive body of knowledge about when to trust what comes through these senses, and how to interpret what is seen.

The difference between this and the case of the microscope is that (natural) eyes are being used in the first case but not in the second. This means the perceptual procedures available in the first case need not be available in the second. Instead the user will have to build up a set of new ones directly by practicing with her new sense-organs. But the new set thus acquired will have the same epistemic status the old ones did: they will be (more or less) independent of empirical science, and all one's evidence gathering will have to be mediated through them.

Therefore, one of my claims about perceptual procedures still holds even if we change our sense-organs: the epistemic justification of the distinctions made by means of our senses operates, largely speaking, independently of scientific theory.

Here's new terminology. Any procedures we use to gather evidence

---

4   See the subsection in Hacking 1981 titled "Truth in Microscopy," it appears also in Hacking 1983: 200–5.
5   By no means am I suggesting that every aspect of learning to distinguish artefacts introduced by a device from what is really out there *must* be handled by mastering gross regularities independent of the empirical sciences. On the contrary, much of this learning may be handled by means of scientific results that tell us what we think we are seeing *cannot* be there, or which provide enough detail about the process we are using to enable us to predict what certain artefacts of the system will be. For a nice illustration of how problems with artefacts can arise in a contemporary research context, see Chaisson 1992.

require *buck-stopping* procedures: there must always be a set of procedures whose epistemic justification is supported by gross regularities, and which are located at our proximate end of any evidence-gathering procedure.

Now, although buck-stopping procedures *are* supported by gross regularities, it seems that the undercutting argument *isn't* true. Namely, it seems *possible* that one could gradually wean oneself from one set of sense-organs, and the senses these confer, and adopt an entirely distinct set of sense-organs with entirely different senses. And this seems to allow *biotechnical incommensurability*, the possibility that all the knowledge gained on the basis of one set of senses (our natural ones, say), and in particular, the scientific knowledge built on the evidence those senses made available, is false by the light of our new set of senses.[6]

I will address this worry more fully in Part II, but this much can be said now. Recall that new sense-organs are ones the operation and implementation of which turns on the recognition of a class of gross regularities, which in turn depends on our senses being able to make certain distinctions. Should we then switch to a reliance on the new sense-organs, they are at least required to underwrite their own operation; that is, they must allow us to see and implement the set of gross regularities needed to operate and fine-tune *them*, and consequently they require a continued capacity to make the (sensory) distinctions these gross regularities rely on that we were able to make earlier. The conclusion is this. The suggestion that we could design new sense-organs for ourselves, and then, as it were, kick away the ladder of our old senses, including the gross regularities and distinctions that we relied on to design these new organs, seems incoherent. We would have simultaneously undercut our capacity to use these new senses.[7]

---

6   "Biotechnical incommensurability" is *not* the incommensurability philosophers of science, in the wake of Kuhn, have worried about. That sort of incommensurability turns on how theories infiltrate the data used to either confirm or disconfirm them, not on worries about modified sense-organs.

7   The details of this argument are fleshed out later (see Part II § 5). One possibility has *not* been ruled out: a radical shift in the capacity of our sense-organs induced by, say, mutation, or the meeting with a species whose sense-organs have capacities which in no sense are continuous with our own. I will not address to what extent radical biological incommensurability arises because of cases of this sort.

§9

CONCLUSION TO PART I

This has been a study in the technology of epistemology.[1] The concern has been with what we *do* to bridge the gap between theory and application: the methods (mathematical and instrumental) that causally connect us to our world. This raises, as we've seen, issues different from those that the traditional linguistic focus is apt to do. That focus, recall, often takes the form of characterizing knowledge as a *successful act of a certain sort*. One therefore tries to augment the notion of justification with that extra something so as to yield knowledge – and failures to know are just so many counterexamples to purported definitions of knowledge.

My concerns have not been with how to formulate a paradigm of knowledge that will never include epistemic failures, but to study how we (in practice) routinely *turn* epistemic failure into (provisional) epistemic success. This is what makes natural a concentration on the scope of scientific programs, and how we extend them, using, among other things, articulated and unarticulated gross regularities. Our sheer complexity (as organisms), and the sheer complexity of the objects and instruments we are in causal proximity to, leads to the notion of perceptual procedures (and, more generally, to the notion of buck-stopping procedures). The crucial properties of perceptual procedures are:

(a) that they are underwritten by both articulated and unarticulated gross regularities which, applied as they must be to ourselves and middle-sized objects around us, are for the most part out of the reach of the scope of any science, and

(b) that for biological reasons they are necessary *whenever* we want to interact with anything in the world.

---

1 It has also been a study in the technology of *reference*, as will become clear in Parts III and IV.

82

Perceptual procedures, at last, provide us with the epistemic foundation for the sciences in terms of what *we do*.[2]

One last thought. It may seem that I have given everything required for a quick argument against Quinean views on revisability. Gross regularities which underpin perceptual procedures *must be* part of our conceptual scheme, so it would seem, for we need them for our perceptual procedures, and, barring mutation or technological advances (and even *with* such things), we will always need something like these perceptual procedures. Thus, however much we revise our conceptual scheme, such gross regularities must be present: they are unrevisable.

This *doesn't* follow, because nothing guarantees what vocabulary the gross regularity is to be expressed in. From the point of view of a class of possible conceptual schemes (which are rigid vis-à-vis our perceptual procedures), I show in Part II that what is true is that gross regularities must be encoded in the truth-conditions of sentences which otherwise express a great deal more: gross regularities must be *part of* the content of (some of) the sentences in our web of belief. But nothing requires that the sentences expressing the gross regularity must be the same (have the same meaning *in toto*) as we go from possible conceptual scheme to possible conceptual scheme. The regularity expressed by "Grass grows from the ground," in one conceptual scheme may be expressible in another only by the statement "The hair of the earth-god is rooted to his scalp." The Quinean letter is intact, but its epistemological spirit is denied.

2   Subject to the caveats raised by the possibility of surgically enhanced sense-organs. I claim (and claim again, when the subject is raised later) that this possibility's impact on epistemology is less than one initially expects.

Part II

TWO-TIERED COHERENTISM

The first most remarkable fact about the intellectual history of our species is the predictive and explanatory success of common-sense middle-sized-object ontology.

(Jerry A. Fodor)

§ 1

INTRODUCTION TO PART II

Most contemporary philosophers think epistemic foundationalism ran aground because it couldn't distinguish theories from data-bearing statements for those theories. The current offer is a holistic continuum between theoreticity and observationality, one dense with sentences in the middle and bereft of them at the end-points. The result has done much for philosophers: An apparently naive picture of scientists inductively gathering incorrigible facts and settling them in a congenial theoretical network has been replaced by a vision of discontinuous mutations of theory/data bundles from one incommensurate form to another.

Although arguments obliterating hope of a theory–observation distinction are strong, the holistic view has problems too. One is this: a cursory glance at scientific practice over the last 300 years reveals enormous *progress*. But, alas, any description of medical advances, household appliances, or transportation innovations cannot be treated by the holist as the indisputable advances they are. There's no denying, it seems, the possibility of scientific change eliminating the claim of any such development to be a scientific "success."

This seems wrong. I argue that conceptual change *requires* deep continuities and allows the accumulation of facts – it's just a matter of seeing where these facts are. Blocking insight has been the focus on our scientific heritage as a scheme of statements. Instead, we must attend to methods of gathering evidence, and the *structural* requirements they place on conceptual change.

Traditional epistemic foundationalism sought its Archimedean points in an incorrigible set of ideas, or, later, in an incorrigible set of statements. This doctrine was given its final form by Carnap: to rebuild our conceptual scheme on a foundation of sense-data statements and analytic truths, where the positivistic proposition serves both as a semantic unit, a meaningful truth-bearing object, and as an epistemic unit, an item open to revisability or incorrigibility. Quine's rejection of this substituted an observational–theoretical *smear* in which the two roles were still merged.

Separating the two roles and assigning them to different vehicles reveals an epistemic rigidity in our evidence-gathering practices masked by the linguistic flexibility of our language.[1]

Sketch of Part II

In §§ 2–3 I present two-tiered coherentism. I start with standard semantic objects: the interpreted sentence with its truth-conditions, the term with its extension, and construct epistemically significant analogues of these. In § 4, I extract the notion of an observational regularity from the apparatus constructed in §§ 2–3, and show that it is the repository of inter-theoretical evidence, and one source of accumulated knowledge in science. In the rest of § 4, and in §§ 5–6, I show that this source of inter-theoretical evidence and acquired fact is not endangered by arguments in favor of incommensurability, and I illustrate other sources of accumulated knowledge in science. Finally, § 7 summarizes and addresses methodological worries about naturalized epistemology.

---

1   I admit it sounds strange to claim the locus of epistemic properties is *not* something senten-
     tial or propositional. Isn't it *sentences*, or something like them, that we believe, and attempt
     to confirm or disconfirm? Nevertheless, I'll show in what follows that progress in episte-
     mology is made by substituting something else for the sentence or proposition as *the*
     significant epistemic unit, at least when it comes to revisability and observationality.

§2

EVIDENTIAL CENTRALITY

Before starting, let me motivate details of the forthcoming construction. The undercutting argument (hereafter *UA*), raised in Part I, shows that certain aspects of our web of belief are required by the machinery of conceptual change; and because whatever is required for the evidential justification of a conceptual change cannot itself be flushed out with the debris, those aspects are unrevisable. I first explored this strategy for unrevisability while investigating one strand in the traditional notion of the *a priori*: to see by means of it whether logic, in whole or part, must be held rigid during conceptual change (see Azzouni 1994).[1]

The variant of the strategy explored here uses the result of Part I that observational practices, as a whole, can't be jettisoned by conceptual change. The question, then, is how the presence of these practices manifests itself in the web of belief. Because of holism we won't see them *at all* if we fasten on to sentences, propositions, or any other meaningful linguistic vehicle. Instead, it is certain observational application conditions of sentences that stubbornly remain with us after any conceptual change. I describe the class of these rigid elements as *evidentially central*.

The UA defines a kind of content – evidentially central content (ECC) – not expressed by interpreted sentences in a way bereft of the other semantic content those sentences have by virtue of belonging to our (or any) web of belief. ECC, though, must be present in any conceptual scheme: there must be sentences which express it. ECC, in contrast to Balaguerian nominalistic

---

1   My conclusion: logical principles themselves need not be held constant during conceptual change, but only certain *application instances* of those principles, that is, those *uses* of them in evidential contexts (see Azzouni 1994 Part 3 § 4). Those who want an argument for the *a priori* status of classical logic – as a whole – won't find it by looking to the evidential tools used to justify theory change. For a contrasting view, see Field (1996), who takes the reliance on standard logic of confirmation theory, probabilistic views of epistemology, and so on, to fix the presence of that logic during conceptual change. I am not convinced by this argument, and my forthcoming construction doesn't treat classical logic as an Archimedean point in conceptual change.

content or van Fraassenian observational content, is not linked to the truth-value of what is expressed, but only to what remains invariant in the face of conceptual change. Thus it doesn't have the problems with truth that these other notions face.

Given conceptual change, what remains invariant? The UA tells us that the application of gross regularities to certain contexts where evidence is evaluated (but not every context in which these regularities are taken to be true) must always be evidentially available. So for each gross regularity used evidentially, I single out those of its truth-conditions which the UA implies must stay with us – and call that set a *naked proposition*. The set of naked propositions – or, more precisely, a set constructed from the set of naked propositions – is *evidentially central* – required in any conceptual scheme (used by us) to gather evidence.

Which gross regularities are used in evidence gathering? Confirmation holism dictates the answer: *all* of them. In contrast, in § 3 of Part II, I single out *observational regularities* from the class of gross regularities. For the moment, the reader can take the success of this move on faith.[2]

Let's begin. The phrases "conceptual scheme," or "web of belief," label a collection of interpreted sentences in a given language $\mathcal{L}$, each paired to a truth-value – this corresponds to the composition of our conceptual scheme by beliefs that certain interpreted sentences are true and others false. I sometimes write of a set of sentences being *in* such a conceptual scheme, and mean by this that a collection of ordered pairs, each one an interpreted sentence paired with the value *true*, is contained in the conceptual scheme.

I presuppose the notion of an (interpreted) sentence: a linguistic item with *truth-conditions:* I assume possible contexts in which meaningful sentences have truth-values. Each context, and resulting truth-value for a sentence, is a *truth-condition* for that sentence.[3] Putting things this way allows me to speak of "changing some of the truth conditions of a sentence," and to mean that some of the truth-values assigned to the sentence – as a linguistic object – in certain contexts, have shifted. The truth conditions of "grass is green" in English makes "grass is green" true where grass is green, and *false* otherwise.

In a classical setting, only two truth-values are admitted, the positive and negative truth-assignments to any sentence disjointly exhaust the set of

---

2   The construction of this section goes through even if the class of observational regularities cannot be singled out from the larger set of gross regularities. In this case, although the evidential center (ECC) of our conceptual scheme will still be definable, its proximity to observation will not have been established.

3   I'm not committed to any particular way of formalizing my talk of "contexts," in terms of situations, possible worlds, or whatever. The use I make of this apparatus is, semantically speaking, so simple that what I say (when restricted to standard two-valued logic) is compatible with the various alternatives.

contexts, and the negation operator on sentences allows us to restrict ourselves to positive truth-conditions. But belief systems governed by non-classical logics (where this isn't true) exist; and the forthcoming construction is neutral on this.[4]

A *definition*: Let Δ and Ξ be two sets of sets of truth-conditions. We say that Δ is *nicely contained in* Ξ if there is a 1–1 mapping Ω of Δ into Ξ so that for each set of truth-conditions TC in Δ, either $TC = \Omega(TC)$ or TC is a subset of $\Omega(TC)$.

An *example*. Consider a frog named Jack, all contexts that include Jack, the sentence "Jack is croaking," and let Ξ be the unit set of the set of truth-conditions of that sentence with respect to the contexts containing Jack; all contexts, that is, where Jack is croaking (on the one hand) paired with the value *true*, and those where Jack is not croaking (on the other) paired with the value *false*.[5] Now consider the set of all contexts in which Jack *and* Jill are present, and let Δ be the unit set of the set of truth-conditions for "Jack is croaking" relative to those contexts. Δ is nicely contained in Ξ.

The meanings of interpreted sentences are often taken, by practitioners of formal semantics, to be fully characterized by truth-conditions (e.g. sets of all possible worlds in which they are true); however, in contrast, the individuation conditions of interpreted sentences are taken by some philosophers to include the logical relations which sentences have to other sentences. For example, it has been thought, and not unreasonably, that two atomic sentences with the same truth-conditions do not have the same meaning if one belongs to a classical system and the other to an intuitionistic one, because they are not logically equivalent to the same sentences.

My project is unaffected by this issue. But I need a term that refers to the associated truth conditions of a sentence, while abstracting away from other factors, e.g. logical relations, that affect the meaning of a statement (relative to a conceptual scheme). Call a set of truth-conditions a *proposition*, and when a proposition is a subset of the truth-conditions of a sentence (in a conceptual scheme), describe that sentence as expressing that proposition – where it is understood that sentences in conceptual schemes will, in general, express more than the proposition(s) they express, and this even when the

---

4   In the general case, the truth-conditions for a statement S are ordered pairs of contexts with truth-conditions, where the truth-values are from among $t_1 ... , t_n$. Truth-assignments are to arbitrary sentences, and not merely to atomic sentences without logical structure. This is deliberate because, in general, the logical structure of an interpreted sentence with the very same set of truth-assignments may vary from one conceptual scheme to another: it can have connectives and quantifiers in one scheme and be atomic in another. This raises the complication that, relative to the particular logic of a conceptual scheme, sentences cannot be assigned truth-conditions independently of each other. I grant this; it won't bear directly on the issues here.

5   This set of truth-values is not exhaustive of the contexts.

proposition expressed is identical to the truth-conditions of the sentence expressing it.[6]

I can now describe what sort of thing is evidentially central. Start with the gross regularities used to execute perceptual procedures, and with the gross regularities that express relations recognized (on the basis of executing these perceptual procedures) as holding between objects, either as these things are expressed in *our* web of belief (in the case of articulated gross regularities) or as how we act on them (in the case of unarticulated gross regularities). I can't use the set of truth-conditions of sentences expressing articulated regularities as part of the set of naked propositions, because these sentences express too much: The terms appearing in them – being terms of *our* conceptual scheme, and because of holism – have truth-conditions apart from the particular situations we can use them in to gather evidence.[7] So, for any sentence expressing a gross regularity, I take only the set of those truth-conditions for it, relative to those contexts where it can (generally speaking) be applied in evidence gathering, to be the naked proposition expressed by that sentence. Doing this for every sentence expressing a gross regularity (used evidentially) provides a set of naked propositions.

An example. You dip litmus paper into a liquid and subsequently see it turn red. To verify this, you rely on the truth-conditions of a number of gross regularities about, among other things, what colors look like in a certain light, and when viewed from certain angles and distances. Were the light in your lab peculiar, you would have stepped into the hall to examine the litmus paper.

The truth-conditions employed are narrow in scope, at least as far as their epistemic necessity for the evidence-gathering procedure just executed. Contrary to fact, it could be the sentences we normally use to claim things about light, colors, distances, and so on, are pretty much false most of the time; this doesn't matter as far as the bit of evidence gathering in question is concerned, provided the truth-conditions of these sentences are intact in the context containing the events that have just taken place in your lab. That is,

---

6   "Proposition" so used is close to notions found in formal semantics, depending on exactly how "context" is defined, e.g. the popular construal of propositions as mappings from possible worlds to truth-values. The important differences are that

    (1) I'm not restricted to classical two-valued logic – and so do not see the "meaning" of an interpreted sentence as necessarily exhausted by the "proposition" it expresses; and

    (2) propositions, on my view, need not be *complete* collections of truth-conditions.

    This means (for me) that any interpreted sentence expresses one maximal proposition as well as all the subsets of that proposition.

7   For example, sentences expressing gross regularities generally hold in situations that are observationally inaccessible to us, or inaccessible altogether.

the truth-conditions of sentences expressing the articulated gross regularities needed for epistemic support here are only a very few of the truth-conditions of the sentence. And it is only these evidentially necessary truth-conditions that belong to naked propositions.

I've been vague about exactly *which* sentences about our visual experiences we rely on to draw conclusions about the liquid being an acid. This is because, in practice, what's relied on are *un*articulated gross regularities, something that should not be surprising given Part I § 4.

It might seem that to speak of truth-conditions for unarticulated gross regularities is to treat them as linguistic, when it's been argued that they're not. *Truth-condition*, however, is a concept with broader powers than this objection recognizes. An unarticulated gross regularity can still be recognized as applicable or not applicable to situations – and this is all that's needed for it to have truth-conditions.

The set of naked propositions constructed via the UA is not yet the required evidentially central class of items; it's a *set of sets* of naked propositions that is evidentially central. Here's why: our evidential practices can survive two sorts of excision from the set of naked propositions expressed by the interpreted sentences of a web of belief, or used in the application of that web. First, our evidential practices can survive the removal of one or another naked proposition we otherwise rely on; second, they can usually survive excision of a small number of truth-conditions from one or more of the naked propositions; in effect, replacing the set of such naked propositions with a set of naked propositions nicely contained in that set. What they cannot survive is excision of *most* of these naked propositions or their truth-conditions. Here is a (somewhat gruesome) analogy: I can manage pretty well if any one skin cell (or even a small clump) is removed from my body. But if *most of them* go, I am in messy trouble.

Notice an implication of the skin-cell analogy that is to be taken seriously: sorites-style arguments do not work here. One cannot get around the requirement, that most of the evidentially central naked propositions and their truth-conditions must be retained, by harmlessly removing them one by one over time. This is true of naked propositions for pretty much the same reason it's true of skin cells: these things serve pragmatic roles. If one or two, or a few, are missing, one can still get by, practically speaking, with the rest. Something awkward can sometimes happen, but it won't cause a general collapse, that is, the sort of general collapse which becomes inevitable when most of them are missing.

Here's another analogy.[8] Consider the various translations of number

---

8 *This* analogy is helpful on condition of an acquaintance with how numbers, and relations among them, are defined in terms of sets, and relations among them. Benacerraf (1965) contains a readable informal version of how this goes.

theory into set theory. The set-theoretical sentences that "1 < 2" is mapped to by means of different translations differ *radically* in their set-theoretical meanings; but the inferential role the sentences play vis-à-vis each of their *translated* fellows, that is, the other number-theoretic statements, is identical. What changes from translation to translation is only the relationships of the translated sentences to other *non-numerical* set-theoretical sentences.[9]

The same point can be made another way. Regard the numerical sentences as partially constrained sentence structures, and their internal parts (the successor function-symbol, "0", etc.) as semantically incomplete structural units. Each set-theoretical translation of the terms of number theory gives different set-theoretical content to each numerical statement structure: one translation may treat "<" as the member-of relation while another treats it as the ancestral of the subset-of relation. Regardless, the inferential roles the statements of number theory have vis-à-vis each other must be left intact. The translations do this by respecting structural restraints placed on the number-theoretic terms.

Here's how this analogy illuminates evidential centrality: Replace "inferential role vis-à-vis other number-theoretic statements" with "epistemic role due to a certain class of evidential practices." And shift the concern from the various set-theoretical definitions possible to the various conceptual schemes possible. A set of naked propositions, contained in the evidentially central set of sets, may nicely embed in the truth-conditions of the interpreted sentences of any number of conceptual schemes, so that what changes is the inter-relation of the images of the evidentially central set of naked propositions (under an embedding) to the interpreted sentences of the conceptual scheme which *do not* express naked propositions,[10] as well as what the terms appearing in the statements (the naked propositions are mapped to under an embedding) refer to. Regardless, the epistemic role of the image of a naked proposition (under an embedding), insofar as it serves as a repositor of a particular naked proposition, is the same.

This, then, is my claim: our epistemic practices reveal an evidentially central set of sets of naked propositions; and one or another of its sets of naked propositions must be embeddable in our conceptual scheme, regardless of how science modifies that scheme. Here's how I prove it: the UA tells us that science, for the most part, cannot overthrow the results of perceptual procedures without undercutting the very methods used to justify scientific claims. It is the way perceptual procedures support the UA which has been milked for the rigid epistemic elements needed to construct the evidentially central set of sets of naked propositions.

---

9 I.e. the statements of set theory which, relative to a translation, have no statements of number theory mapped to them.

10 As I observed (note 4), this may be due to the statements (in the different conceptual schemes), expressing the embedded propositions, having very different logical structures.

The construction is fairly abstract; in this respect it's analogous to existence proofs in mathematics. I've used the UA to prove the existence of an evidentially central set of sets of naked propositions – and I've also indicated by an example what naked propositions are apt to look like. But this tells us very little; only a more detailed examination of the gross regularities employed in evidential contexts will tell us more.

This brings me to my next point. I've stressed the use of the UA in this construction, but the independence condition for gross regularities, too, plays an important role. For I first considered sentences that express gross regularities, and it was natural to do this because of the independence from empirical science of perceptual procedures, and our reliance on gross regularities to execute them. Only then did the UA further whittle down the truth-conditions of the sentences expressing gross regularities, and thus provide naked propositions.

Because I restricted myself to gross regularities, I did not capture *all* the naked propositions that belong to the evidentially central set of sets of propositions. To get *those*, a construction close to the one carried out here must be executed on *all* the sentences of a conceptual scheme used in evidential contexts. Such a construction, not restricted to sentences expressing gross regularities about observations (as this one was), but including, say, simple logical and mathematical principles also used in evidential practices, provides, in addition, the naked propositions expressed by logical and mathematical truths.[11]

The construction described in this section relies on applying the UA and the independence condition to the interpreted sentences of *our* conceptual scheme. Why, one might worry, should the evidentially central set of sets of naked propositions be the same one if it is constructed from the vantage-point of some other conceptual scheme – especially one with a different logic (and especially since a different logic can call for additional truth-values)?

I've assumed that implementing an interpreted sentence in a conceptual scheme, or the rejection of such implementation, are fundamental practices that cannot be given up.[12] Thus nice inclusions of two-valued naked propositions are not ruled out by changes to non-two-valued conceptual schemes. The UA, then, tells us that any conceptual scheme that can be adopted by us is one in which a nice embedding of one of the evidentially central sets of naked propositions exists, because gross regularities themselves provide distinctions that survive the replacement of those regularities by other gross regularities, should our perceptual procedures be replaced by other

---

11 This is the point made in Azzouni (1994 Part 3 § 4).
12 This is behind the idea, in the study of multi-valued logics, of designated truth-values, in terms of which consequence relations are defined. See e.g. Urquhart (1986).

(buck-stopping) ones. This doesn't tell us that the construction yields the same evidentially central set of sets of naked propositions if it is carried out on a radically alien conceptual scheme – i.e. not one we can see ourselves adopting. For such schemes, as far as what I've said here is concerned, anything goes.

Naked propositions are constructed from statements by stripping away some of their truth-conditions. Remaining at the truth-value level, however, prevents sight of other epistemically significant structure, the discovery of which turns on a close examination of the terms occurring in the sentences that express gross regularities. An indication of this is my having left the relation between a sentence expressing a gross regularity, and the gross regularity itself, unexplicated. In the case of observational gross regularities, there is something to say about this; but we must look at terms to see *what*. I expose the epistemic structure in noun phrases in a way analogous to the construction of naked propositions: Just as the naked proposition is a subset of the truth-conditions of a statement, so too there is something epistemically significant corresponding to the term to be found after removing many of the instances a term holds of. This is the job of the next section.

§3

OB-SIMILAR EXTENSIONS
AND OB*-SIMILAR
EXTENSIONS

I made the humdrum observation earlier that we connect our (empirical) terms to their referents by means of *procedures*. We can recognize sea water by taste: this procedure (tasting a liquid and getting a certain result) is associated with "sea water." We have also more subtle and complicated methods by which to recognize the presence of sub-atomic particles associated with the names for such particles.

Here the suggestion naturally arises that philosophers once used to make the above commonplace do philosophical work: there's a distinct class of commonsense terms with their own "ordinary" criteria of application. That is, we recognize what these terms apply to merely by means of our biologically given senses, and without using theoretically justified instruments.[1]

That certain procedures, rooted ultimately in our commonsense practices, pick out the extensions of commonsense terms, is most plausible in respect of terms for foodstuffs, certain kinds (e.g. copper, gold, and water), certain types of animal (e.g. cats), for colors, and so on. However, Putnam[2] has argued forcefully that language is a cooperative venture. No lay-person can rationally ignore an expert who points out that what he is about to drink, despite his best efforts at identifying it, is *not* water, or that the ring he's purchased for his fiancée is not gold. For one way science advances is by justifying new procedures for picking out items in the extensions of our terms; and, *to some extent*, by faulting old ones. The ancient Greeks could not recognize the presence of iron or gold in either stars or (certain) chemical compounds, and it would have been possible to fool them, in ways that would not fool us today, about the presence of iron or gold in substances.

This means that attempts to isolate a class of commonsense terms with

---

1   Actually this is, at best, a rational reconstruction of what certain "ordinary-language" philosophers were up to. The sorts of distinction they tried to make were varied in nature, and none of them, I believe, *quite fit* the distinction I am trying out here, one designed with the discussion of procedures in Part 1 § 7 in mind.

2   (1975a: 227–9), for example, and elsewhere.

their own (perceptually based) procedures will fail. Indeed, terms like "water," "gold," and "sulfur," already seem part of chemistry, and consequently of physics (despite their ancient pedigree), and terms like "human," "tiger," and "leaf," seem to fit well with biological nomenclature. So it looks like commonsense terms *already* belong to one or another science (and consequently the hope of marking out commonsense terms which operate independently of the sciences is doomed to failure).

This is too fast. As other philosophers have pointed out,[3] "gold," "water," and other substance terms, as used by ordinary folk, tolerate impurities in their extensions in scientifically unprincipled ways (so, for example, "soda" is not taken to be "water," as the terms are ordinarily used, although "sea water" is). Similarly, ordinary terms for kinds of living being ("dog," "mouse," "ant," etc.) *do not* mark out natural kinds from the biological point of view.[4] Finally, there are many terms ("airplane," "car," "chair," etc.) which belong to no science, and are *at best* functionally definable. One may *apply* aerodynamics *to* airplanes, but "airplane" is hardly an aerodynamical term.

This won't help the ordinary-language philosopher yearning for special procedures for such terms, at least as ordinary folk employ them. For nothing about commonsense usage of substance terms and biological-kind terms precludes scientific methods from being applied to them to develop new procedures for recognizing instances of such commonsense "kinds"; more strongly, nothing (except intractability) precludes outright definition of such terms in scientific nomenclature. And as the case of "airplane" makes clear, this is true also of artificial-kind terms. There's no reason to think that commonsense terms (even if not definable in scientific terms) have isolated criteria of application.

But we are not without hope of salvaging *something* wanted by the ordinary-language philosopher, since not all procedures are alike. Recall the distinction (Part I § 7) between perceptual procedures and instrument-enhanced procedures. Although I've argued that this distinction is fairly solid when it comes to procedures, it won't straightforwardly lead to an analogous distinction for terms. One might have hoped, that is, for a naturalized division between observational terms and theoretical terms. "Observation terms" would be those terms with only perceptual procedures associated with them; the rest would be "theoretical terms" (on the grounds that any non-conservative instrument-enhanced procedure would have to be, in part, justified on the basis of scientific theory).

This seems to face problems for two reasons. The first is that just about *any* simple-kind term (nowadays, anyway) has non-conservative instrument-enhanced procedures associated with it. This is true even of color terms:

---

3   E.g. Wilson (1982). Nagel (1961: 7) makes the point in passing.
4   See Dupré (1981) for detailed discussion of this, and for interesting examples.

there are times when looking *to see* the color of something will not work (because, say, it's emitting concomitant radiation which makes it impossible to view directly). But a device that detects light radiation, and emits coded sound patterns in response to such radiation, gives us instrumental access to its color.

The second reason is that not every (non-conservative) instrument-enhanced procedure need be backed up by scientific theory, as the extended discussion of gross regularities in Part I indicated. If we sort coconuts into two kinds ("soft" and "hard") because a certain kind of rock can crack one but not the other, no theory is involved.

Surprisingly, all is not lost; but I need new terminology to show this. The *extension* of a procedure (associated with a term) is the set of things the procedure recognizes as falling under that term. Its *anti-extension* is the set of things the procedure recognizes as not falling under the term. The considerations thus far show that no extension of any natural-kind term coincides with the extension of any perceptual procedure.

Some observations about the extensions and anti-extensions of procedures

Generally, the extension and the anti-extension of a procedure do not exhaust everything there is: procedures yield partially defined functions. There are always limitations to what a procedure can be applied to – for as we've seen, causal connections between ourselves and things are not easy to construct or exploit. Testing for gold by checking for ductility, for instance, will not verify either the presence *or* absence of gold in our sun; and particular chemical methods for recognizing gold compounds are restricted in applicability to those gold compounds. Finally, procedures are *fallible*. The method for recognizing the presence of gold by checking the substance for ductility, yellowness, etc., gives the wrong answer when applied to *sea water*.[5]

Although differences among procedures don't lead to a straightforward distinction among terms, there is still something to learn about vocabulary by attending to procedures: it seems to be a triviality that some differences between things can be recognized by the naked senses, and some can't. I use "ob-similar sort" ("observationally similar sort") to name a maximal collection of items or events that can be classed together by means of a set of perceptual procedures. There are commonsense terms which *very nearly* mark out ob-similar sorts: terms for foodstuffs, "tomato," "potato," "spaghetti"; terms for certain plants, "dandelion," "oak," "grass"; and most

---

5   In part, for the reason just given, no extension of any natural-kind term can be said to coincide with the extension of any procedure or family of procedures, perceptual or otherwise, as I show in Part IV.

terms for furniture. There are also commonsense terms that very nearly mark out ob-similar sorts for events: "lightning," "wind," etc.

The reason for the qualification "very nearly" is that it's always possible to fool the senses with an event or item which doesn't fall under the term in question (so that the extension of the class of perceptual procedures used to pick out the extension of a term does not exactly coincide with it).[6]

The set of ob-similar sorts can be refined indefinitely by means of tools into a set of sorts which have, epistemically, pretty much the same status the original sorts have. The "tools" here are ob-similar stuffs brought to bear against each other. For example, two ob-similar sorts of rock, A-rock and B-rock, may be divided into further (disjoint) subclasses A_1, A_2, B_1, B_2, because A-rocks in one subclass (A_1) may be broken apart by B-rocks in another subclass (B_1), although the remaining A-rocks (A_2) are too strong; the rest of the B-rocks (B_2), meanwhile, can break any A-rock. Depending on the point of view, either B-rocks instrumentally sort A-rocks, or vice versa.

Call the resulting collection *ob*-similar sorts*. This set properly contains the set of ob-similar sorts, and any procedure for an ob*-similar sort is a finite string of procedures for ob-similar sorts applied *in a series* to one or more ob-similar sorts as the basis. The previous illustration relied on two relevant ob-similar events: A-rocks-crumbling-when-struck-by-B-rocks and A-rocks-failing-to-crumble-when-struck-by-B-rocks, recognition of which we apply one after the other to mark out ob*-similar sorts.

Although instrument-enhanced procedures are required to recognize ob*-similar sorts, these are still crude distinctions: we can *define* the class of ob*-similar sorts in terms of ob-similar ones. But, most crucially, what makes these distinctions crude is the absence of (scientific) *theory*. If refining these distinctions *ad infinitum* were all there is to scientific practice, then everything we need to do science would have been essentially stated in Mill's methods.

Ob*-similar sorts are nothing more than divisions among ob-similar sorts. But the really powerful thrusts in science provide the underlying *identities* among things. A major advance was the identification in the seventeenth century of various phenomena (e.g. lightning, the amber effect, the electric eel) as "electrical." Ontological postulating with accompanying theory is unavoidable here – and, in general, there is no definitional route from ob-similar extensions to the theoretical entities needed, or the distinctions and identifications resulting because of them. Witness viruses, sub-atomic particles, gravity, and the large numbers of events explained and unified in terms of these things.

---

6 Some will disagree with this, especially those who hope for a law-like connection between such terms and what they refer to. Such a connection, if it existed, could be used to define a set of procedures that would exactly pick out the set in question. In Part IV I give arguments against this possibility.

An illustration. Chemical theory distinguishes those cases where something (an element, say) is present in another form and distilled out (by one chemical means or another) from other cases where something (a chemical compound, say) is actually created by a process. No refinement of ob-similar sorts gets us this distinction.

More generally, scientific theory explains how otherwise heterogenous-looking procedures linked by a term (e.g. "gold") actually pick out the same thing; there's no way to achieve this with the simple refinement of ob-similar sorts by the application of procedures to distinguish items among them.

Ob- and ob*-similar sorts are the maximal natural units available to us through what we can reasonably call "observation." Because every *natural-kind* term has non-conservative instrument-enhanced procedures associated with it, the *extensions* of natural-kind terms are structured.

A definition. A *core* of a natural-kind term is any ob*-similar sort that is (for the most part) contained in it.[7] The indefinite article is needed because terms can have more than one core: Both the set of non-microscopic diamonds and fairly large manifestations of ash are cores of "carbon." Also, many natural-kind terms have no cores: "positron" and "chromosome" are examples.

I call terms with cores *core terms.* Core terms include the terms for middle-sized objects we interact with on a daily basis, but don't include every term commonly used. "Black hole," "atom," and "virus" are not core terms. On the other hand, many terms for particular kinds of plant and animal ("epiphyte," "adelomyrmex ant"), detailed names of hardware parts ("pilot bit," "hex nut"), names of kinds of rock ("coquina," "diorite"), names of *many* elements, and so on, *are* core terms, although they need not be commonly used.

I've made a distinction in vocabulary, although not one useful for distinguishing commonsense terminology from the rest of our vocabulary. It won't support a class of incorrigible truths; it does its epistemic work not by isolating "observational terms," but rather by highlighting terms whose extensions contain something epistemically special (i.e. cores).

That the extension of no natural-kind term is coextensive with an ob*-similar sort indicates the parochial nature (ontologically speaking) of ob*-similar sorts. This is why I've avoided "kind" in respect to ob*-similarity and used the neutral and informal-sounding "sort" instead. Let me dwell on this a moment.

---

7   In calling these ob*-similar sorts "observational," it is not, recall, being suggested that we are infallible in our attributions of terms to items. The clause "for the most part," thus, is there because any procedure yields false positives. Also, the scare quotes around the term "natural kind" (which I am not consistent about using) indicate that such terms need not be "natural kinds" in the strict scientific sense.

Ob*-similar sorts are a heterogeneous bunch vis-à-vis the natural kinds of the sciences. They usually fit into the natural kinds of one science or another, if at all, only by being split up into gerrymandered parts. This is notoriously true of ob-similar color sorts (the red things grouped by eye have rather disparate physical sources; and the same is true of colored items even if we restrict ourselves to wavelengths, as the phenomenon of metamers makes clear).[8] The same is true of ob*-similar types of animal and plant. The problem is that any ob*-similar sort is a simple instrumentally induced subset of an ob-similar sort. But any sequence of procedures used to pick out the instances of such a sort is fallible; and often the distinctions thus made aren't fine enough in view of subsequent science (and, I might add, in light of more subtle instrumental procedures).

Second, as mentioned, even when an ob*-similar sort fits neatly into a scientific kind, it rarely extends the breadth of that kind: For science often proceeds by identifying instances of a kind that cannot be recognized by the procedures previously used to recognize instances of it.

That ob*-similar sorts, scientifically speaking, are not ontologically stable, means we cannot predict how the instances of an ob*-similar sort will sort into our scientific worldview (provided, of course, they sort into it at all). There was no telling ahead of time that what ancient Greeks picked out as gold would (largely) remain one kind of thing, and there was no telling ahead of time what other apparently different stuffs would turn out (in some sense) to be gold (or to contain gold) as well.[9]

Here's a nice illustration of how ontologically slippery core terms can be. Consider ordinary talk of images or appearances. Images are not sense-data, since one can be mistaken about what they look like. On the other hand, talk of images is neither merely talk of the objects they are images of nor, for that matter, simply talk of the properties of the viewer. One may use what one sees in the telescope lens to recognize properties of Uranus; but it is equally common to successfully refer to, and describe the properties of, the image without describing or intending to describe the planet at all. But then, ontologically speaking, what *are* images?

Luckily, I don't have to answer *that* question, although philosophers of perception have worried about it for decades. I need say only that this difficult question is a special case of the general ontological problem ob*-similar sorts pose. Talk of images is evidentially necessary; that's why talk of them can't be eliminated. How an image looks in a telescope may reveal how the

---

8 See Hardin (1988).

9 Ferromagnetism is a good example of how radical things can be. Magnets seemed to be a kind of stuff: stone with unusual powers. Well, in some sense that's right, since ferromagnetic molecules are special. But the magnetic force exhibited by the lodestone is a ubiquitous phenomenon, best understood as a relativistic manifestation of electricity (or, better, something we might call *electromagnetism*).

telescope should be modified (a lens in it cleaned, shifted, etc.). On the other hand, it's clear that a simple neat ontological category for images isn't forthcoming: the image I see through a telescope, no matter how similar to the image I see while on LSD, cannot be ontologically identical to it.[10] What images are, ontologically speaking, shifts from context to context, and can sometimes prove quite obscure.

So much for the elusive ontological qualities of ob*-similar sorts. The moral is this: *when we can bring the sciences to bear on them*, they are revealed to be deeply parochial. From the point of view of the *programs* of the sciences, ob*-similar sorts can't participate in the ultimate limning of the universe. But we are stuck with them anyway for epistemological reasons. They are projections, as it were, of regularities detectable by the human sensory system; that is, they are the groupings of items into sets of things *we* can distinguish – and *all* our procedures for evidence gathering turn on our ability to distinguish them. The existence of ob*-similar sorts, and our reliance on them, reflects our need for perceptual procedures to gather evidence.

These last remarks may seem to go too far by taking us beyond ob-similarity to ob*-similarity, and so I'd better sort things out a little. Recall yet again that perceptual procedures operate (more or less) independently of the sciences, and that we rely on them for evidence gathering. The distinctions that perceptual procedures can make for us are precisely the ob-similar sorts, and so the remarks one paragraph back certainly apply to them.

What about ob*-similar sorts? Their epistemic position is pretty similar: Since they are recognized by bringing ob-similar sorts to bear on each other in an instrumental capacity, they raise no new epistemic issues. Scientific theory is not involved, and so no dependence on empirical science arises; furthermore, since perceptual procedures are fallible to begin with, ob*-similar sorts bring no new epistemic risks. It is hard, actually, to see how to separate ob-similar and ob*-similar sorts epistemically.

The epistemic rigidity of ob- and ob*-similar sorts explains an otherwise puzzling phenomenon about ordinary usage. Despite massive scientific development, most commonly used core vocabulary remains curiously fixed, at least as far as their cores are concerned. Consider, as examples, terms like "light," "water," "fire," "grass," "air," "gold," "fish," "up," and "down." As strikingly, such terms seem to pose no problems when translating them from one scientific culture to another – again, provided their cores are focused on. Semantic holists can try to explain these facts by suggesting that the terms are remote from the locus of scientific change; but this won't do for most of the terms mentioned above. Scientific change has impacted on them, and continues to do so, directly and radically. Another (and more popular) holist

---

10   Well, philosophers *have* disagreed with this; but they haven't had an easy time as a result.

strategy is to claim that the terms so translated *are* different, and that we are deluding ourselves in thinking such terms, before and after scientific changes of the magnitude contemplated here, are decently inter-translatable.[11]

My explanation is different. Ob-similar sorts must stay with us because they're the only distinctions our senses are capable of making directly; and thus the cores of any term are epistemically central. Indeed, an examination of the above terms, and others like them, shows that certain cores for such terms that are central from the point of view of general usage have remained the same despite far-reaching changes involving the introduction of new cores, new procedures to pick out relatively esoteric instances of the terms, and/or (sometimes radical) shifts in the theories the terms are embedded in. The class of solid, relatively pure, and sufficiently large samples of gold and iron are unchanged cores of "gold" and "iron" since Antiquity. Despite radical changes in beliefs (the theories, mythological or otherwise) about wind, the perceptual procedures (the sensations) for recognizing wind are unchanged.

Here's another example.[12] Consider the commonsense "up" and "down." Our view of what up and down come to has changed much since Greek times, for we take these terms (cosmically speaking) to be far more localized than the ancient Greeks realized. Nevertheless, as far as ordinary usage and our evidence gathering procedures are concerned, nothing has changed: we use the same (kinesthetic) method to tell which way is up and which down that they did, and our results are the same.

The differences between professionals and the average individual with respect to core terminology arise in how and when someone can be mistaken about a term's application. Leaving aside the unusual mistake, everyone, pretty much, successfully picks out items from the cores of core terminology. Scientists, however, also successfully pick out what a term refers to in broader domains connected to their specialities.

Pressure exerted on a term from its cores explains the points made by philosophers such as Mark Wilson (1982), cited earlier. A core, although epistemically natural, may not fit comfortably among the natural kinds that science urges upon us. Soda and sea water are alike enough in percentages of water and impurities as to make it hard to justify treating soda as not water, in contrast to sea water, as people ordinarily do. Such strains in usage are relatively minor matters linguistically, and the evidence for this is that, despite the slippage between terms like "water" as used scientifically and as used by the rest of us, scientists shift easily between professional jargon gained later in life, and the lingo used at home; there, they rarely experience

---

11 At his most militant (e.g. 1962), Feyerabend takes this view. See also Churchland (1979).
12 I owe the example to Feyerabend (1962: 85–6), though my discussion starkly contrasts with his. Field (1989: 11) offers pretty much the same picture as Feyerabend.

a desire to impose reforms on lay-relatives because ob-similar sorts play such an ineliminable role in observation, and thus in ordinary ways of classification. This explains how we can use core terminology to pick out ob- and ob*-similar sorts, even though, strictly speaking, the references of core terms are not restricted to such, and may only partially overlap with them.

Let's return to the issue raised at the beginning of this section: whether terms can be defined which have only perceptual procedures associated with the items in their extensions. Ob*-similar kinds are precisely what is wanted, although they *do not* correspond to ordinary-kind terms. This is because they require the explicit description of the procedures that pick them out as part of their definitions. "Red" won't do, but "looks red when procedure *p* is applied" will. Generally, the way to capture an ob*-similar kind definitionally is to give the scientific kind(s) the ob*-similar extension is largely contained in, and pick it out of the kind(s) explicitly by mentioning whatever parameters the substance must fall within to be detectable by the senses, or by mentioning whatever procedures must be applied to something to identify it. Sometimes, scientific kinds need not be mentioned at all: just the procedures involved for recognizing instances will do. The *definitions* available by using scientific-kind terms or instrumental procedures shifts as scientific jargon does, of course (this is the cost of such terms being theory-laden); but we will always, with one definition or another, be able to pick out the same ob*-similar sorts.[13]

I want to connect ob*-similar sorts with the construction of the evidentially central set of sets of naked propositions from § 2. Recall that sentences expressing gross regularities used in evidential activities were operated on by the UA to produce the evidentially central set. But I left open the precise nature of the relation between the sentences expressing such gross regularities and the gross regularities themselves, except to indicate vaguely that the gross regularities in question are required to execute perceptual procedures.

I can now say more. The gross regularities in question are precisely about ob*-similar sorts, and so the sentences that express them are ones with core terms. I call these gross regularities *observational regularities*, and I say more about them in § 4. For now, I'll stress that sentences expressing observational regularities can shift as scientific theory shifts because the terms in them pick out more than their cores, and this *more* depends on scientific theory. The cores, however, are rigid because of the independence condition: they are the natural observational items that arise from our perceptual proce-

13 This discharges the promissory note from Part I § 5, note 17. Ob*-similar terms are terms that are theoretically laden, but use only perceptual procedures for the identification of what they refer to (in order to be seen as terms that pick out only "observational" entities). Of course my remarks about the parochial nature of ob*-similar kinds are not in the spirit of van Fraassen.

dures, and which, because of intractability, are more or less epistemically independent of the sciences. Since they are natural epistemic units because of the independence condition, and not because of UA, gross regularities about cores are still broader than what's needed, strictly speaking, for evidence gathering. (And so this is why UA was used in § 2 to whittle down gross regularities to naked propositions.)

Two last points.

(a) It is the independence condition, as I show in § 4, applied to perceptual procedures, that allows the accumulation of what we might call Baconian facts (observational regularities) about the world around us, and which contribute to the growth of scientific knowledge.

(b) The UA relies on the existence of perceptual procedures (or buck-stopping procedures), and so it is narrowly restricted in its application to observational regularities. But the independence condition applies to gross regularities generally, regardless of whether they are observational, and this blurs (§ 5), in one significant sense, the epistemic differences between observations of the world and theoretically justified instrumental interactions with it. The impact of the independence condition on our epistemic practices is more important to scientists interested in data that can bear against theory in a way that is independent of theory. The UA is of more interest to philosophers implicitly concerned with the issues of revisability and scepticism. Certainly that explains my interest in it.

§4

OB-SIMILARITY, OBSERVATIONAL REGULARITIES, REASONS FOR INCOMMENSURABILITY

Some philosophers suggest that there's an interesting relation between laws and natural kinds.[1] To have a collection of laws is to implicitly have a group of natural kinds that obey the laws. Conversely, to have a collection of natural kinds is to implicitly have a set of laws they obey. However things sort out for laws and natural kinds, this relation holds of ob*-similar sorts and ob*-regularities (or, for naturalness, "observational regularities").

Ob-regularities are gross regularities holding among ob-similar sorts (and similarly, for ob*-regularities in respect of ob*-similar sorts). Indeed these regularities are just what are used to build ob*-similarity out of ob-similarity. Ob*-regularities have the same epistemic status that ob*-similar sorts do, although actually establishing this fact via perceptual procedures, as I did with ob*-similarity, is pointless since ob*-regularities are gross regularities, and these have already been established as (more or less) independent of science.

The next point to make about ob*-regularities is that, since the precise terms picking out ob*-similar sorts can shift, so can the language describing regularities associated with such sorts. This shift in language is usually linked to the scientific fate of the regularity: the ob*-similar sort(s) the regularity holds of may become more remote from the natural cleavages of the universe science reveals, and the scope of the regularity may be similarly restricted. What starts out looking like a fundamental law may in time become an illustration of a quite local phenomenon not deserving of the rubric "law" at all; nevertheless, the regularity will (more or less) still be honored (although not in the original terms).

Suppose we have terms $\mathbf{A}$[2] and $\mathbf{B}$ and the regularity:

(†) $\mathbf{B}$, generally speaking, follows $\mathbf{A}$

---

1   See e.g. Nagel (1961: 75).
2   A temporary convention: I use bold variables ($\mathbf{A}$, $\mathbf{B}$, $\mathbf{C}$ ...) for terms, and non-bold variables (A, B, C ...) for the sorts such terms to refer to.

Imagine that the procedures associated to date with **A** and **B** only pick out items of which (†) holds. Several things can happen. First, **A** and **B** can come to have new procedures associated with them, so that they're taken to hold of more than they were taken to hold of originally. In this case, (†) may no longer hold of A and B but only of (scientifically uninteresting) subclasses of A and B. Second, we may decide that **A** and **B** don't really pick out anything special at all: (†) is an accidental generalization based on several coincidental but not otherwise related phenomena. Third, we may find subclasses of A and B that a real law holds of (perhaps **A** and **B** as originally used are sloppy: they pick out too much).

And other cases are possible.[3] But in each one a linguistic shift (motivated, admittedly, by scientific progress) has disguised that what had been established still holds: it just cannot be expressed the same way any longer.[4] If the special cases are epistemically significant to us, we will still be aware of them. Although we know (or most of us do, anyway) that the *sun* does not rise and set, the old regularities about sunsets and shadows hold nevertheless; it is the terms these regularities must be expressed in that are very different.

Consider again "up" and "down." The regularities detectable by the senses and associated with these terms still hold (what goes up must, generally speaking, still come down). What has changed is how we explain these regularities and how far we take their scopes to extend.

Let me recapitulate. As we shift from web of belief to web of belief, we must take our perceptual procedures with us. Consequently, any web we adopt must be anchored at the periphery, varying Quine's image, in ob*-similar sorts. As we've seen, *it doesn't* follow (and anyway, is practically never true) that we must have scientific-kind terms that pick out the ob*-similar sorts. Nevertheless

(1) we *can* always pick them out by alluding explicitly to the procedures used to recognize their presence;
(2) they play a central epistemic role in our practices; and
(3) the regularities they support (or which support them) *must be held true* for the most part, no matter *how* they are expressed.

I've mentioned several times that core terminology is not theory-neutral. In saying this, I am *not* holding hands with the holist, for I mean this in a fairly

---

3   (†) could end up epiphenomenal – to be explained as the reflection of some *other* regularity about *other* kinds entirely.

4   A beautiful example of a collection of regularities that eventually could not be expressed scientifically (because the *kind* term they were couched in vanished) are Davy's observations about *Conferva rivularis*. *Conferva rivularis* subsequently turned out not to be a natural kind (see Hacking 1983: 152–4).

weak way. First, I am willing to concede that beings with the same biology as ours can have a language that cuts up cores *somewhat* differently than ours does (i.e. culture has *some* impact on core vocabulary: recall the soda example mentioned in § 3). Second, I have conceded that science can often modify the procedures associated with core terms, and can sometimes modify the scope of regularities. And, third, I concede that the programs of the sciences give us new beliefs about the items referred to by core vocabulary ("chairs contain quarks").

Despite the limitations placed on the theoretical permeability of the *cores* in core vocabulary, it's clear why nothing like a traditional observational–theoretical distinction can be drawn: Cores are hardly the *whole* story about the *references* of core terms: they are, at best, only the whole story about the *sensory* relevance of core terms.

One important constraint on evidence gathering in science that falls out of what I've argued for here, and in Part I § 7, is the *observed regularity constraint* (ORC): *almost all observational regularities constrain theory construction in the sciences.*

Despite the considerations supporting ORC, it still might seem to fit badly with the history of science. For this is a history of theoretical *change*, and some find a close examination of it to show theoretical incommensurability.[5] One source for incommensurability is taken to arise from the fact that two competing traditions in science can disagree on the *relevant* data. That is, they can disagree on how regularities should be described, and worse, can disagree even on which regularities are relevant to the debate. But I've argued for a largely rigid set of observable regularities. Must I reject this commonly cited *datum* for incommensurability?[6]

Well, *no*, as it turns out. Theoreticians often toss out regularities. And no doubt competing theoreticians can toss out *different* ones. But this is not to ignore the regularities.

First, there's a long road from regularities detectable by the senses to scientific theory. And there are many ways (in principle) to explain how an apparently incompatible regularity actually fits with a theory. Such explanations range from postulating theoretical objects (a hitherto undiscovered planet) exerting certain effects on the phenomenon, to various ways the regularity could be an evidential glitch (mechanism malfunctioning in a systematic way, human bias, and so on).[7]

---

5 See, among many others, the writings of Feyerabend, Kuhn, Lakatos, and Lauden.
6 I won't commit myself as being either for or against (various sorts of) incommensurability. There are certainly periods when it is unclear which scientific theory, if any, should be adopted. But I find it debatable whether this shows the deeper claims incommensurability proponents have argued for. The *datum* I've mentioned I am convinced of, however, and so I need to show its compatibility with largely fixed observable regularities.
7 Recall the discussion of empirically measured deviations from a theory in Part I §§ 4–5.

Let me expand this point in respect of experimentation. It's often noted that one successfully conducted experiment can count for far more than hundreds of failed ones. Why? One reason, surely, is the competence of the experimenter. We don't take failed high-school experiments seriously, and not because high-school students are missing a sociological badge of authority which admits them into the cadre of those whose failures should carry weight. Rather, it's that high-school students are (generally) not skilled at executing experiments, and their lack of skill and resulting sloppiness can be detailed by showing which botched (most likely unarticulated) gross regularity was needed to carry out a particular experiment.[8]

Failed experiments by highly skilled practitioners are also disregarded, at times, because of the learning curve involved in the mastery of new apparatus (or newly designed apparatus).[9] Even when both experimenter and apparatus are quite seasoned, results can be disregarded because of an awareness of what can go wrong. On the other hand, a positive result, if repeatable, may be seen as so unlikely that it is taken seriously despite failures in similar experiments. This is because of the recognition, just mentioned, that something else may be complicating the failed experiments.[10] In any case, both failures *and* successes, even systematic ones, may be disregarded for an indefinite amount of time if experimenters cannot make themselves reasonably sure that they have managed to design the experiment to exclude (within certain thresholds) all extraneous complications.[11] Of course, both repeatable failed experiments as well as repeatable successful experiments elicit observational regularities. The issue for the experimenter is how to explain the regularity (not ignore it). If the experiment is successful, this is because the best explanation for the regularity turns on whatever it was the experimenter was testing for or studying. Otherwise, the regularity is due to something else, and an experimenter may be reasonably convinced of this even if he cannot find out exactly what.

Another way of eliminating apparently disconfirming regularities, by the

---

8   No doubt the rather low quality of the equipment they have to work with plays a role, too.
9   See the description of Bob Harrison's attempt to operate a laser in Collins (1985: 58–72). Not once did he take his inability to get the laser working as positive confirmation, however small, that laser technology of this sort is impossible.
10   George Smith has pointed out to me that Galileo's inclined plane experiments, used to show the acceleration of motion due to gravity, give results as much as 28 percent off if the angle of inclination of the plane is too great (because of complications from the difference between rolling and sliding, a matter not sorted out until Euler over a century later). Does this mean that the inclined plane results, even where they give (more or less) the expected results, are a failure? Not at all. A successful result at the angles of inclination Galileo actually measured is enough of a surprising (and robust) result to allow us to draw the (provisional) conclusion that a hitherto unrecognized complication is flawing the other results, rather than that the successful ones are bogus.
11   For an interesting case study illustrating these points, and others, see Franklin (1993).

way, is by denying they're the sort of thing the laws of the particular science apply to (at least in any obviously direct way).[12] This *doesn't* mean denying the regularity *altogether*. Rather, it means moving it to the domain of another (scientific) department. If it proves intractable or too much of a special case for any branch of science to handle, it *will* be ignored (as something to be *explained* – we may *use* it evidentially *anyway*, as gross regularities indicate). This may not seem respectable, but any scientist faced with such a fact should suspect the theories and tools at his disposal can explain it – concern with it is a waste of time otherwise. Thus, "constrain" as used in ORC is fairly weak: scientific work is always work in progress, and this means there can be regularities, or anomalies, that remain unexplained *forever*.

To sum up: perceptual procedures, and the ob*-similar sorts that can be generated by their means, function as an epistemic foundation for the sciences: the data available through them cannot be eliminated, for the most part. Nevertheless, this foundation is compatible with a datum often given in support of incommensurability: that perceived regularities may be treated in different ways by scientists in different evidential traditions.

12 That mammals stay roughly the same temperature despite (relatively small) changes in the temperature around them is not explained by the *direct* application of principles of thermodynamics.

There are no general criteria for recognizing which regularities are relevant to (or tractable for) a particular science and which ones aren't. Sometimes the only way to tell is for a few scientists to spend time trying to explain something a certain way.

§5

KUHNIAN CONSIDERATIONS AND THE ACCUMULATION OF KNOWLEDGE

Theory has other ways of invading observation besides tainting the words the observations are couched in. There is the psychological approach, a theory affecting the *observer* so that she *sees* (in the most fundamental way we understand this verb) things differently after adopting it. Kuhn, in particular, sometimes *seems* to go this way by denying the existence of regularities verifiable through the senses. Rather, *gestalt* shifts in perception allow one to "see" regularities where one did not see them before. And so, "the proponents of competing paradigms practice their trades in different worlds."[1]

The examples Kuhn gives to illustrate this thesis, however, are compatible with the existence of observable regularities since, first, that one might need training to *see* observable regularities is not puzzling provided we don't squeeze observation into the Procrustian bed of sense-data. Second, scientific observations usually involve instruments and, as mentioned earlier, developing new technology to study something out of our naked sensory reach invariably means laboriously discovering *new* observable regularities. These new regularities usually amount to the fact that when a device designed in such and such a way is operated in such and such a fashion a certain result usually occurs.[2] However, early versions of devices rarely give clearly observable regularities and, consequently, it's hard to use them to tell

---

1   Kuhn (1970: 150). Similar language is in Collins (1985).
2   "Laboriously," for the literature makes clear it's not easy to design instruments which operate in a regular manner (so that data may be extracted from them). See, apart from (already-cited) works of Hacking and Collins, the writings of Feyerabend, Hanson, Kuhn, and others.

       I should add that most of this literature (excluding Hacking and Collins) is concerned with how theory-dependent instrument-development is. Kuhn (1961) is an excellent example. Nevertheless, my point *stands*. For theory here is a method of *refining* measuring techniques: it points out the direction in which refinement is supposed to take us. Nevertheless, the resulting instrumental procedures must be (more or less) *autonomous*. That is, operating them should give us regular results (i.e. observable regularities), otherwise the results won't, in turn, exert pressure on the theory (and precipitate crises). None of the discussion in this extensive literature shows this requirement to be violated.

what properties the objects being studied have. For example, Kuhn (1970: 115–17) notes that, prior to Herschel's actual discovery of Uranus, there were at least seventeen different occasions on which a star was seen in the position where Uranus (we now suppose) must have been. To explain this, we need not invoke a cultural *gestalt* shift that allowed viewers to reclassify what they had been seeing all along. Rather, unsurprisingly, since Uranus is so far away, someone gazing at it via a primitive telescope will have a hard time seeing *anything* clearly enough to draw conclusions, not to mention observing the item clearly and long enough to see something "odd," about it.[3]

Kuhn's other examples can be handled similarly, and so I skip the details. It's instructive, though, to compare Kuhn's views with those of Gould (1981). Gould wonders how scientists managed to collect evidence for the claim that differences in the size of skull cavities are linked to race and sex, when we no longer can find any such evidence. Gould implicitly takes ORC as a given,[4] and thus can't speak merely of paradigm shifts or perceptual changes, and leave it at that. Rather, he repeats the experiment himself, and looks for places wherein bias can insert itself – places, among others (including miscalculations, etc.), where the crudeness of the experimental technique prevents the emergence of observable regularities theoretically connected to what is being measured. He finds them: in one case, mustard

---

3   Hoskin (1995: 171–3) describes the Uranus episode in a way that supports the independence of observation from theory:

> That evening [Herschel] was studying stars in Taurus, when he came across one that – so excellent was his mirror, and so experienced was he as an observer – he instantly recognized as anomalous: "In the quartile near ζ Tauri the lowest of the two is a curious either Nebulous Star or perhaps a Comet ...".

Hoskin observes (ibid.: 173): "Interestingly, it did not occur to Herschel that his 'curious' object might be a planet, perhaps because ... he was unaware of the widespread opinion among astronomers that there were ... planets to be found." Hoskin notes that professional astronomers initially had trouble locating the object Herschel was referring to,

> for whereas the organist with his home-made reflector had seen at a glance that this was no ordinary star, the Astronomer Royal with the professionally-built instruments of Greenwich Observatory had been able to find nothing unusual in that region of the sky and had been forced to identify Herschel's object by its movement.

Hoskin observes (ibid.: 174) that "the unknown amateur's discovery of a planet moving so slowly that its change of position in a single night was almost imperceptible, astonished astronomers throughout Europe and gave notice that an observer of exceptional ability had appeared on the scene."

4   Although what Gould *does* presupposes ORC, when he describes his aims (1981: 27), he sounds like Kuhn: "If – as I believe I have shown – quantitative data are as subject to cultural constraint as any other aspect of science, then they have no special claim upon final truth." Unfortunately, showing that bias can generate *mistakes* in the collection and measurement of data (which is *all* that Gould does) implies no such grand claim.

seed was used to measure the volume inside a skull, and different ways of pressing the seed into a skull can lead to significantly different measurements of volume for skulls of (nearly enough) the same volume.

It is only when one fails to locate where the original experimental setup could have fallen short (of producing observable regularities) that one should consider saying with Kuhn (1970: 135) that "the data themselves had changed." And such replication is never easy. One must duplicate the cramped workspace, the flickering candlelight, *something* of the mindset, the crude instrumentation, and so on, to determine to what extent measuring techniques fail to provide observable regularities.[5]

As I've mentioned before, science seems to accumulate truths, although work in the philosophy of science in the last thirty years or so disparages this claim. But if there are observable regularities which are fixed despite conceptual change, as I have argued, the accumulation of knowledge of such regularities constitutes precisely a kind of accumulation that can be credited to science; for applications of science, to the extent that they reach observable objects, offer new observable regularities. To take only two simple cases, the building of an atomic bomb or a washing machine can be described in terms purely of ob*-similar sorts and observational regularities. And the techniques (supported by such regularities and sorts) for designing these products remain, regardless of how scientific theory subsequently shifts.

This suggestion of what is cumulative about science does not require the fixing of the standards, the terminology, or the styles of reasoning of the sciences. And, it places the accumulation of truths precisely where the layperson sees it so patently – in applications.[6]

It is worth pointing out that *anything* we *do* in the sciences (or anywhere, for that matter) can be described in terms of a recipe composed entirely of perceptual procedures. This follows from the simple fact, first mentioned in Part I, that any procedure we employ must contain perceptual procedures at our end of the causal chain. Such descriptions have two grave drawbacks, however:

(a) They must be very long, for anything not recognizable directly via perceptual procedures must be described in terms of its pedigree: the recipe for the perceptual procedures used to make it from perceptually recognizable stuffs must be given.

---

5   Hacking (1983: 177) has a charming example of how tough this can be. Herschel conducted a number of experiments on radiant heat using filters. One filter, apparently, was a nearly black brandy in a decanter. *This* filter is no longer available.
6   Having used Kuhn as a stalking-horse, I must add that he sometimes expresses sentiments somewhat different from those attributed to him here. He writes (1977: 339) that "proponents of different theories can exhibit to each other, not always easily, the concrete technical results achievable by those who practice within each theory."

(b) Such descriptions do not provide *explanations* of why the recipe takes the form it does. Such explanations usually turn on theoretically postulated entities, the mentioning of which is excluded if descriptions in terms only of perceptual procedures are allowed. What is given, that is, is simply a "brute" recipe. That is why descriptions of scientific experiments are *never* given purely in terms of perceptual procedures.

But they *could* be. And so whatever knowledge is encapsulated in such recipes remains, regardless of how scientific theories were used to derive them, or how such theories subsequently change. And this is because of the (relative) epistemic independence of perceptual procedures from science.[7]

Two other points about the accumulation of scientific knowledge should be raised. The first is that it would seem that what's required of the accumulation of observational regularities as knowledge is not only that the number of such items increases, but that their scientific status is *(relatively) stable*. Should a gross regularity be explained by being embedded among scientific laws in some way – perhaps because the ob*-similar kinds it is about are themselves hedged into genuine scientific kinds – we would like that status to remain. There is the impression that science progresses not only by generating observational regularities (about refrigerators, say) but also in the stability of the explanation of the regularities, when they are to be had.

Second, another source of apparent stability in science, not explained by the mere accumulation of observational regularities, has to do with the *status* of certain entities and facts about them. It's been pointed out by more than one philosopher (and scientist!) that belief in electrons seems legitimately more entrenched than does belief in the theories about those items. Theories about electrons might not remain in the form we now have them, but no one believes that electrons themselves are in any danger from this. Does anything I've had to say in the foregoing bear on this claim, one way or the other?

Yes, and I can link the two kinds of stability in scientific theory development with one sort of explanation. Perceptual procedures are (more or less) independent of scientific theory, and because of this they generate a body of gross regularities about objects we perceive. The important point is that we can learn about these objects in a way that is (more or less) independent of our beliefs and theories about those objects. (This is why good observations can lead to surprises.)

Consider electrons again: if belief in them is more robust than are the

---

7   Of course, such knowledge can be lost: the apprentice experimental tradition can be interrupted; the particular apparatus or treatment of materials that enables a certain regularity to occur can fail to be passed on to others. But these sorts of loss do not bear on the independence of such knowledge from scientific theory.

theories about them, then, in some sense, our methods of finding out about them must be independent, to some extent, of our theories. This can't be true in the same strong sense that it's true of our perceptual methods, but it can be true in a relative sense.

How? Well, consider the unobservable (theoretical) entities of science, items such as viruses, electrons, quarks, and so on. Theory allows instrumental interactions with them. If we can forge instrumental relations between ourselves and a class of objects so that

(1) the results of operating these instruments are independent (epistemically speaking) of what the operator of them believes;
(2) we have means of adjusting and refining our instrumental access to the things being detected;
(3) our instrumental access to things enables us to track properties of them (either in the sense of detecting what they do over time or in the sense of taking time to explore different aspects of them); and
(4) (certain) properties of the objects can be used to explain how it is we know (possibly other) properties of them;

then we have achieved thick epistemic access to them.[8]

Scientific theory enables us to understand how thick epistemic access is acquired: exactly what sorts of causal transduction are taking imperceptible properties of non-observational items to perceivable events. Nevertheless, conditions (1)–(4) allow the recognition of gross regularities about unobservables: we can discover patterns of behavior in items like quarks or electrons that we cannot explain scientifically.[9] Such gross regularities are (sometimes temporarily), therefore, independent of scientific theory in this sense, although not in the sense that they would survive a general collapse of scientific doctrine altogether – since then our reasons for thinking we even *had* thick epistemic access to something would cease.

Gross regularities so established may, in turn, connect the theoretical entities they are about to certain observational regularities in a way that scientific theory proper has not. For properties of electrons discovered via thick epistemic access to them can lead to empirical applications; such

---

8   These conditions on thick epistemic access are drawn (with modifications) from Azzouni (1997a).
9   There are many examples of this, especially in the study of sub-atomic particles. We are able to formulate theories in which neutrinos have mass, and theories in which they do not. It is thick epistemic access to them which tells us which theories among these are to be taken seriously. See, for a popular exposition of this, Kearns *et. al.* (1999), which contains an accessible discussion of both the theoretical issues *and* the particular instrumental interventions to learn how neutrinos "oscillate" between flavors. Another good example of trying to learn about particles through thick epistemic access is the search for proton decay.

applications are explained by properties of electrons so discovered, although they may not be explained by one or another *theory* in physics. (We may know that electrons act in such and such a way and exploit this behavior of theirs for other purposes without being able to explain, in terms of our fundamental – currently believed – theories of electrons *why* they act this way.)[10]

The upshot is that the gross regularities about a theoretical entity may connect it to observational regularities in more ways than even well-established theories about it do, and so such entities can survive scientific change (up to a point) because the gross regularities utilized in applications have an explanatory role we want to preserve.

Ultimately, therefore, it is conservativeness that makes the electron more robust than the scientific theories around it: the more applications gross regularities about electrons are employed in, the more essential those gross regularities (and the entities – electrons – they are about) become to our overall web of belief.

I promised that the explanation on offer here would also explain the stability – to the extent there is such – of the well-established scientific explanations we have for observational regularities; indeed, it would explain what "well-established" comes to. This explanation is the flip-side of the claims just made about theoretical gross regularities: as I've said, many explanations of observational regularities arise not from austere scientific theory but from an intermediate body of theoretical gross regularities, which are thus better entrenched with regard to observational regularities than scientific theory itself. This means that we are, if we can be, conservative with respect to our explanations of observational regularities in terms of theoretical gross regularities. None of this *must* survive scientific change: there is no UA to take advantage of; there is only an independence condition supplied by thick epistemic access in a way analogous to how perception supplies an independence condition for observational regularities.

10 Accompanying such theoretical gross regularities, of course, can be concepts that fail to be "definable" in the terms of more pristine science. An example – from chemistry – is provided by the various theoretical gross regularities about the "shapes" of molecules.

§6

PERCEPTUAL
IMPERMEABILITY AND
BIOTECHNICAL
INCOMMENSURABILITY

I have not yet responded *directly* to the possibility that what we perceive can be infiltrated by the categories of our language and our beliefs (that perception is *cognitively penetrated*); I've only claimed that the examples given by Kuhn and others from the history and practice of science do not *compel* such a view. But isn't infiltration possible? And, if so, then change in doctrine *could* change what is perceived: a change in our scientific beliefs could make ob-regularities literally disappear (or appear) before our eyes, obviating the independence of perceptual procedures from the sciences.

Furthermore, it might seem that neurophysiology and the related philosophical literature are where to look to resolve this issue. Unfortunately, there are two problems. First, what's known of neurophysiology doesn't help. Second, in any case, much of the related literature there is tied up with something irrelevant, the question of what impact neurophysiologically innate response-categories have on language. Consider color as an illustration. Hardin (1988: xxii) argues that "the semantics of ordinary color terms is powerfully constrained by the physiology of the human visual system." Van Brakel (1993) disagrees, contesting the neurophysiological and anthropological literature Hardin's claim is based on.[1]

I am, however, concerned with possible constraints in the opposite direction: from language, or theory, *to* perception; and there doesn't seem to be much around that resolves this issue *conclusively* one way or another. The discussion in Fodor (1983: 64–86) – with which I *am* sympathetic – offers what it can in the way of argument and pertinent psychological literature to show that perceptual systems are modular and, in particular, "informationally encapsulated" (p. 66), that is, that they do not use "high-level expectations or beliefs." His argument involves things like the

---

1 Although I've officially avoided the question of how semantically variable human languages can be, it should be clear that my position allows *great* variability in this respect. This is implied by my separation of the rigid epistemic elements of a conceptual scheme from the semantic ones.

persistence of optical illusions (despite knowledge that one is in the presence of an optical illusion), evolutionary and computational considerations (creatures who see what's out there instead of what they expect or hope for generally survive longer; creatures who deliberate too long to determine what they see don't do well in the long run), and such like, but although convincing, and backed up by the literature available at the time he wrote, is largely speculative.

Saunders and van Brakel 1988 describe a study of the Dani (a group in New Guinea) by Rosch that suggests that color terminology in language impedes or enhances the capacity to learn new terms. But this bears little on whether perceptual procedures are isolated from the rest of a conceptual scheme and, in particular, from the categories of a language: the isolation of perceptual procedures from the rest of a conceptual scheme does not imply language cannot make it difficult to recognize ob-similar sorts, especially if language supports habitual recognition of certain ob-similar sorts, and not others.

I've been worrying about the scientific literature bearing on my specific claim that perceptual procedures are independent of psychological beliefs, and, unsurprisingly, given how little we know about brains and sensory processes, the result has been somewhat inconclusive. But, look: if perceptual procedures are not independent of our general beliefs and expectations, we should be able to detect this without sophisticated scientific results from neurophysiology. *For recognition of it should already be present in our ordinary evidential practices.*

To some extent, therefore, perceptual procedures' independence from the sciences has already been argued for by pointing out that the Kuhnian suggestion that ob-regularities disappear when scientific beliefs change is not supported either by standard scientific perception of the source of scientific disagreements, or by our subsequent (historical) approaches to analyzing the source of these disagreements.

The same is true of ordinary evidence gathering outside science. There is a striking difference between how we approach what is to be seen, heard, smelt, etc., where it is the result of perceptual procedures, and where it isn't. In the first sort of case, we point to, or draw the person's attention to, distinctions in the phenomena she should take note of.[2] In the second kind of case – say, when explaining that an acquaintance actually is not angry –

---

2   Part I § 6 shows that these distinctions need not be expressible in language: if I teach someone to apply joint compound to drywall, I'll show him how (at what angle) to hold the taping knife, and how to move and shift it over the wall to sense deviations in the surface, and correct for them. This is all done fairly *non-verbally* (how could I do it otherwise?).

    Even a little work with machinery, e.g. the repair of car engines, or in carpentry, tiling, and so on, does much to dispel the impression that the distinctions made with one's senses are dependent on beliefs or expectations.

we rely on collateral information, and may speak explicitly of cultural differences.

Let's turn to another issue. I've offered an argument for ORC that turns, in part, on our biological limitations. But, although still currently in the realm of science fiction, recall from Part I § 8 that it's logically possible for us to *change* these limitations by replacing our eyes and ears with new artificial sense-organs. Not only could we perhaps subsequently detect new observable regularities, but others might literally vanish from sight, and with them the perceptual procedures that underwrite certain ob-similar sorts.

I'm going to use this possibility to show that ORC requires much weaker premises than those previously used to support it. Those premises did double duty: they supported claims about constraints on *evidential procedures*, and they supported claims about constraints on *evidence*. But, as it turns out, even if, as in the imaginary situation just described, the constraints on *evidential procedures* change, that doesn't affect the constraints on *evidence*. I go on to establish this claim.

Despite our new sense-organs we, presumably, can still recognize that beings with old organs classify items around them in certain ways. Call a classification scheme *robust* if:

(a) the scheme is free of conspiracy (the beings make the same distinctions on individual cases, for the most part, without having to confer with each other); and
(b) it is indefinite (the number of items the beings can classify with it is open-ended).

Robust classification schemes are third-person definable in our own terms: namely, as what beings (of such and such a type) recognize as belonging together (in such and such a way). Furthermore, a robust classification scheme needs explanation. The explanation *can* take the form of a subjective limitation of the senses of such beings. But this won't evade the point: By shifting on our sense-organs, we do not escape ob-similar sorts defined by means of those senses, nor do we escape regularities so detected; we only shift procedures for recognizing them from procedures *we* carry out to procedures *someone else* carries out. We *may* escape, in other words, the need to employ such regularities in our evidential procedures, but ORC is not concerned with *that*. It demands that the regularities in question (in the fullness of time) be explained, and that's something we have not escaped, regardless of *who* or *what* observes them. What is doing the work for ORC is the independence condition, not the UA.[3]

---

3   This means that the gross regularities about theoretical entities discovered by means of thick epistemic access also require explanation – perhaps in the fullness of time. Which is *just* how they're generally regarded in science.

In a sense, that old insight sense-data theorists almost had has finally been captured: they hoped to establish a class of truths without epistemic risk. The mistake was to coax such truths into a canonical form in a canonical language. What is true is that, no matter how our language twists and turns, there are regularities, expressible *this* way or *that* way, as truths, but which are always expressible. Part of what reveals this is that such regularities need explanations, although the kinds of explanation they receive are coded in the way they can be expressed in a web of belief. Perhaps a particular regularity of this sort is an artefact of the sense-organs; in this case, the form it takes in the web of belief will be very different from the form it takes if it marks out something "really" there. But there's no getting away from these regularities.

Not quite, that is. Something that looks like a robust classification scheme can fail to be one. But two points are relevant to this concession. The first point I've already made: if something looks like a robust classification scheme but isn't, *that* it isn't can be recognized without importing the heavy artillery of science. The second point is more important: despite the hopes of eliminative materialists (as expressed, for example, in Churchland 1979), we rarely escape observable regularities by shedding terminology.

Here's a notorious example. The belief in witches has been quite widespread (and is still to be encountered in *some* circles). *We* no longer believe in such beings: we eschew the entire nomenclature of demonic phenomena. But it doesn't follow that persons marked out as witches had *nothing* in common. On the contrary. Procedures for recognizing witches were (and are) based on the recognition of certain kinds of social deviance: "witch" is a term that, when restricted to particular societies at particular times, is part of a robust classification scheme. As the example illustrates, to eliminate a kind term from our vocabulary rarely eliminates our epistemic obligations towards the cores (the ob*-similar sorts) largely contained within that term's scope.[4]

One last observation. In showing that ORC is not dependent on claims about the need for perceptual procedures in evidence gathering, it may seem that I've prevented ORC from restraining science altogether. For, suppose perception *is* strongly influenced by cultural factors: surely an explanation

---

4   I'm skipping lightly over some rather complex sociological territory. If something is a robust classification scheme, that doesn't mean it's not in part the product of culture, as the "witch" example makes clear. In the latter kind of case, the procedures associated with a term can shift from cultural group to cultural group, and shift suddenly within a group (e.g. *what* is taken as socially deviant by a group can shift). Sometimes, in episodes of cultural stress, the application of a term can break entirely away from the procedures usually (in a particular society) associated with it. For an illustration of the latter in Salem, Massachusetts, see Weisman (1984).

of a group seeing a particular regularity, which would be compatible with ORC, *could* be simple cultural bias?

The response to this is really simple: cultural bias alone will not result in a *robust* classification scheme because it relies on conspiracy. Without conspiracy, and without powerful constraints on how one can extend the classification scheme beyond the samples it's (initially) based on, individuals will deviate significantly in how they apply the classification scheme to new cases.

§ 7

METHODOLOGICAL
OBSERVATIONS ABOUT
EPISTEMOLOGY, SCEPTICISM,
AND TRUTH

I've (almost) finished my sketch of the epistemology of the empirical sciences. The rest of the book is concerned with what may, broadly speaking, be called the metaphysics of empirical science. Before plunging into all that, however, I want to give an overview of the preceding, and address methodological worries.

My choice of terminology ("procedural foundationalism," "two-tiered coherentism") plays fast and loose with coherentism and foundationalism, two notions with major roles in recent epistemology. One may wonder where the position designed here lies in the geography of that terminology. For orientation, I describe the standard views, and contrast this one with them.

Epistemic foundationalism requires a class of basic beliefs, or statements,[1] in terms of which all other beliefs or statements are justified. Different properties are attributed to basic beliefs, depending on the sort of foundationalism being pushed. In the Cartesian tradition, such beliefs are seen as infallible, something we cannot be wrong about. Sometimes they're also seen as containing only sensory content.[2] But weaker positions have been popular recently, those which keep the requirement that basic beliefs contain (near enough) only sensory content, while relaxing the constraint of infallibility. Quine is tempted by such a position, hoping by means of his observational conditionals to capture its appealing qualities in his own terms.

Coherentism, in contrast, designates a group of positions rejecting the foundationalist picture just sketched. Justification arises from the coherence of the entire group of statements involved.[3] Generally, no statement is basic.

---

1   It makes a difference to the ultimate position whether the significant unit is the belief or the statement. But it doesn't matter here, and so I'll shift from one term to the other depending on ease of use.
2   I am leaving aside consideration of *a priori* truths.
3   What, exactly, coherence amounts to isn't obvious. Usually it is a sort of logical unity among statements. My own brand of coherence augments standard logical considerations with compatibility in application. I discuss it shortly.

At its most extreme ("pure coherentism"), perceptual statements have no special epistemic status: they may be rejected if they do not cohere comfortably with the body of beliefs already held.

My view blends both positions. Foundationalist elements exist because perceptual procedures are needed in evidence gathering. These procedures are sensory in nature, and capture much of the foundationalist's insight that sensory perception is foundational to evidence gathering. Further foundationalist elements arise from the expressibility of a set of naked propositions, the constraining role of ORC in evidence gathering, and from coherentist epistemic principles operating (for the most part) not with regard to the whole class of statements we believe, but within the borders of a two-tiered structure, where the (largely autonomous) lower tier contains those sentences expressing gross regularities about the results of perceptual procedures. Where coherence patterns (e.g. logical constraints) cross our entire web of beliefs, statements expressing observational regularities cannot be treated with insouciance: The independence of perceptual procedures from the sciences, the UA, and ORC prevent this.

Epistemic foundations thus far, but no further: other traditional foundationalist claims, such as the infallibility of observational statements and the claim that the justification of all statements derives from the justification of statements containing sensory content, are rejected. Since the referents of empirical terms are items in the world, what we take these referents to be shifts with conceptual change. Thus, the Quinean insight that, in principle, revision can strike anywhere, is honored for empirical statements:[4] Any statement may be revised because of scientific development, and so justification of *statements* does not track from a sensory ground-level up. Perceptual procedures are not incorrigible either. We can be wrong about any regularity we think we have observed through misattention, bias, and the other problems observation is heir to.

Coherentism thus far, but no further: without explicitly legislating on the (sticky) question of whether scientific theories are (generally) incommensurable, my position exploits the gap between scientific theory and perceptual procedures in a foundational way to allow for scientific progress. When we step back and observe the broader evidential practices underlying our claims, it's clear observable regularities, as a group, exert a totalitarian grip on justification. This means that regularities describing ob*-similar sorts and observable regularities grow monotonically; and as the considerations outlined in § 6 make clear, this holds in some manner even if the biological constraints on our senses change. The overwhelming impression that scientific practice leads to the growth of *knowledge* is thus explained.[5]

---

4   It's honored for all statements. See Azzouni (1994: Part 3 § 4).
5   I've been illustrating compromises between coherentism and foundationalism. My position

Having said these positive things on behalf of my view, here's a worry: why is this descriptive-looking project *epistemology*?[6] There are a number of ways to answer this question, and in order to justify the project, I'll explore several.

Let's put the question this way: why are the results of perceptual procedures so important for evidence gathering? Why couldn't we, for example, eliminate these procedures and substitute others? For example: start with an arbitrary collection of statements in some language, and add sentences consistent with the set – when we can recognize this – with our (ultimate) target being a maximally consistent set. (Call this *Henkin epistemology*.) What, epistemically speaking, is wrong with this?

Here are some *answers*. *Our* methods track truth better than the alternatives. This isn't a good answer because the truth-idiom can be tailored to fit with any set of sentences we adopt. Mere use of "true" does not place epistemic constraints on the set of sentences adopted.[7] A second answer is that it's conceptually necessary for epistemic practices to be accountable to sensory input. But this is pretty clearly false: we can easily contemplate possible situations where the use of sensory input is evidentially dangerous – for instance, in worlds in which ESP is *far* more solid than sensory evidence. A third answer is that beings with epistemic methods significantly different from ours in respect of sensory input do not survive as well. The problem with *this* answer is that, even if it's right, it's not clear what *survival* has to do with truth. (Maybe our world is demonized: believing falsehoods allows creatures to survive better than does believing truths.) A last answer is that we want to affect the world around us, and our approach enables us to do this in the best way. Unfortunately, this answer begs the question against alternatives.[8]

That no good answer is forthcoming suggests that there's no good *epistemic* justification for the importance of perceptual procedures, in the sense,

---

compromises in another sense as well: It honors Putnam's insight about the linguistic division of labor, while simultaneously acknowledging something of what ordinary-language theorists were groping for when they tried to show that ordinary language is epistemically autonomous.

6 That *any* descriptive project must fail to be epistemology because of its failure to engage with the normativity of epistemology has been raised by, among others, Kim (1988) and Putnam (1983d).

7 I elaborate on this significant point shortly.

8 Interestingly, every answer but the third is precisely the sort of answer one would give to someone, a sophisticated child say, who asked why we try to know things the way we do. One's tempted to borrow Stroud's way of describing the situation, and say there are "internal" and "external" ways of asking the question posed here (see Stroud 1981, 1984). I do *not* adopt this intriguing suggestion.

at least, that justification is being asked for here.[9] Here's why this is true. That epistemic idioms have, intuitively speaking, a normative element associated with them (we're concerned not with the methods we *do* use to know things, but rather with the methods we *should* use to know things) derives from epistemic idioms being *truth*-related (however "knowledge" is defined, it must imply the truth of what is known). There is something special about the truth-idiom; and the intuitions of normativity associated with the epistemic idioms (e.g. knowledge and justification) follow from their relation to "true."[10]

So I must again discuss the truth-idiom, in order to show its impact on epistemic idioms. This excursion into truth also provides an anticipation of the analysis of reference in Parts III and IV, since similar issues arise for both idioms.

Some background. Tarski (1956) placed a famous adequacy condition, "convention T," on any characterization of truth for a language, L: one can derive from it all instances of the schema

(T) "P" is true if and only if P

where the first instance of "P" is replaced by any sentence of L, and the second instance by a translation of that instance in the meta-language. It has been recognized that not just any set of axioms yielding (T) provides the disquotational role that truth is needed for.[11] Call "deflationism" the view that (T), plus additional axioms solely for the purpose of securing the disquotational role, suffices for a description of the truth-idiom and its role in our web of beliefs. Call "minimalism" the view that the disquotational role of the truth-predicate is only a neutral core compatible with various

---

9   Have I shown no good answer is forthcoming just by briefly considering four possible answers? I think so, for it has been made clear what response can be made to *any* answer of this kind.

10 If both "knowledge" and "justification" are normative, and intuitions of normativity are derived from something special about the truth-idiom, why do Gettier puzzles plague attempts to define knowledge as justified true belief? When analyzing a normative noun-phrase *P*, the right question to ask is: *under what conditions can a purported instance of* P *be rejected as a genuine instance of* P? If such conditions differ for two particular noun-phrases, they cannot be defined in terms of each other. In this case, for example, we're not willing to *deny* that a truth-seeking method is justified unless a better alternative is available. Thus the results of methods that sometimes yield wrong answers are regarded as justified even if they're not true. But a belief is *not* acceptable as an instance of knowledge if it isn't true. The normativity intuitions associated with "knowledge" and "justification" derive from the close links of these idioms with "true"; but the links are not the same, and Gettier counter-examples exploit this difference.

For contrasting sentiments, the claim, in fact, that the normativity of knowledge is inherited from its link with justification, see Kim (1988: 218–19).

11 See e.g. Heck (1997).

augmentations, which yield alternative notions of truth. Candidate augmentations are: "verified under ideal conditions of inquiry,"[12] "cohering best with our antecedent set of beliefs," "corresponding to the facts, " or "being justified by some state of information and remaining *justified* no matter how that state of information might be enlarged upon or improved."[13]

"Deflationism," may be attributed to Horwich (1990), though he makes other claims I am not concerned with. He calls his position "minimalism"; Wright (1992) calls *his* position "Minimalism." (They don't have the same position.) What I've called "minimalism" may be attributed to Wright. It also *sounds* like Tarski's view (1944: 362):

> [W]e may accept the semantic conception of truth without giving up any epistemological attitude we may have had; we may remain naive realists, critical realists or idealists, empiricists or metaphysicians – whatever we were before. The semantic conception is completely neutral toward all these issues.

However, Wright (1992: Chapter 1) thinks pure disquotational truth is not a candidate conception of truth; pure deflationism is unstable, *precisely because of convention T*; the notion of truth must be inflated beyond its disquotational role. Tarski's description of other alternatives as "epistemological," however, suggests that *he* would not accept this view; and the statement quoted above seems to even imply that the other "epistemological attitudes" are *not* augmentations of the notion of *truth*. In any case, I disagree with Wright, and show that truth cannot be inflated the way he wants: the inflated candidate conceptions of truth – to the extent that they go beyond disquotation – are purely epistemological in content – just as Tarski suggests. This fact, however, is independent of the disquotational role of truth – no examination of truth's disquotational role tells us, one way or the other, whether it can be inflated or not. This is obvious, if only because axiomatizations of the truth-predicate sufficing it for disquotational use do not by themselves rule out supplementations of those axiomatizations.[14]

Intuitions that indicate how the truth-idiom is ordinarily used show that

(i) disquotational conditions on the truth predicate cannot be supplemented with (non-disquotational) conditions; and

---

12 Putnam (1978a: 1). This is not to say that Putnam endorses the definition.
13 Wright (1992: 47–8) calls this "superassertability."
14 They do not rule out, for example, embedding number theory, with a Tarskian truth-predicate, *non-conservatively* in set theory. Wright (1992: Chapter 1) tries to show that convention T itself forces a distinction between, say, the warranted assertibility of a sentence, *A*, and the truth of that sentence; but the argument turns on a semantic-ascent confusion.

(ii) the truth predicate cannot be *replaced* wholesale by a notion that fails to be a simple disquotational device.

I call this result *the elusivity of truth*.[15]

The elusivity of truth is stronger than the topic-neutrality of truth (Part I § 5). Topic-neutrality requires only that "true" treats the contents of our web of belief the same way; it is compatible, that is, with an inflated notion of truth – provided that notion treats everything necessary to our web of belief and its application the same way; and so it is compatible with the existence of different inflated notions of truth holding of discourses strictly isolated from each other.[16] The elusivity of truth denies the inflation of a truth-predicate, even if the inflated conditions apply to every sentence in the domain of discourse that "true" applies to.

Suppose we try to replace truth with warranted assertability. This fails because of the following "might" statements, which we intuitively feel to be true (I call these statements *Putnam modals*):[17]

(1*a*) "The rug is green" might be warrantedly assertible, even though the rug is not green.

(1*b*) "The rug is green" might not be warrantedly assertible, even though the rug is green.

(1*a,b*) tells us that *warranted assertibility* fails the Tarski adequacy test. If instead we try to introduce warranted assertibility as *either* a necessary or a sufficient condition on truth, we fall foul of this pair of Putnam modals:

(1*c*) "The rug is green" might be warrantedly assertible even though "the rug is green" is not true.

(1*d*) "The rug is green" might not be warrantedly assertible even though "the rug is green" is true.

Putnam modals can be used this way to disable many candidate notions of

---

15 The elusivity of truth somewhat resembles Rorty's views on truth (see Rorty 1986, especially p. 128). Putnam (1978a: 37) should be credited with the forthcoming modals, which show the elusivity of truth, and for seeing how that property of our intuitive understanding of truth is independent of what Tarski calls the semantic conception of truth.

16 Wright (1992: Chapter 2) recognizes this by denying that diverse notions of truth can range over the same discourse, but he overlooks how much this concession infirms the relevance of his investigations of inflated truth for issues of "realism": Since "true" is crucial to applications, mathematics cannot be an isolated discourse in this sense. Neither, I suspect, can moral discourse. Fiction, however, can have its own truth-predicate.

17 See Putnam (1978c: 107–9). Rorty (1986: 127), endorsing the procedure, writes: "'it might be true but not *X*' is always sensible, no matter what one substitutes for *X* ...". I qualify this claim presently.

inflated truth. Consider the constraint of best cohering with our background beliefs, or the constraint of the community of knowers we belong to eventually accepting it as true. These come to grief because of the following modals:

(2a) "The rug is green" might best cohere with our background beliefs even though the rug is not green.

(2b) "The rug is green" might not best cohere with our background beliefs even though the rug is green.

(2c) "The rug is green" might best cohere with our background beliefs even though "the rug is green" is not true.

(2d) "The rug is green" might not best cohere with our background beliefs even though "the rug is green" is true.

(3a) "The rug is green" might be a statement that the community of knowers we belong to eventually accepts as true even though the rug is not green.

(3b) "The rug is green" might not be a statement that the community of knowers we belong to eventually accepts as true even though the rug is green.

(3c) "The rug is green" might be a statement that the community of knowers we belong to eventually accepts as true even though "the rug is green" is not true.

(3d) "The rug is green" might not be a statement that the community of knowers we belong to eventually accepts as true even though "the rug is green" is true.

It seems any attempt to augment, constrain, or define truth that intuitively goes beyond the disquotational role of the truth-predicate fails because of a Putnam modal.[18] Consider, as a test of this, the following eight modals.

(4a) "The rug is green" might be verifiable under ideal circumstances of inquiry even though the rug is not green.

(4b) "The rug is green" might not be verifiable under ideal circumstances of inquiry even though the rug is green.

18 Wright's superassertability fails too: "the rug is green" might remain justified no matter what additional information we get, even though the rug is not green; "the rug is green" might not remain justified no matter what additional information we get, even though the rug is green; "the rug is green" might remain justified no matter what additional information we get, even though "the rug is green" is not true; "the rug is green" might not remain justified no matter what additional information we get, even though "the rug is green" is true.

(4c) "The rug is green" might be verifiable under ideal circumstances of inquiry even though "the rug is green" is not true.

(4d) "The rug is green" might not be verifiable under ideal circumstances of inquiry even though "the rug is green" is true.

(5a) "The rug is green" might fit the facts even though the rug is not green.

(5b) "The rug is green" might not fit the facts even though the rug is green.

(5c) "The rug is green" might fit the facts even though "the rug is green" is not true.

(5d) "The rug is green" might not fit the facts even though "the rug is green" is true.

There are two possible responses to the (4)s. The first reads "ideal circumstances" as having done everything possible to put ourselves in the very best position to know whether the rug is green. In such a case, using elementary possibilities of how mistakes can arise (it isn't necessary, that is, to go *so far* as to threaten the interlocutor with Cartesian scepticism), one can elicit intuitive support for (4a–d). If, however,"ideal" is taken to *rule out* the possibility of a mistake, then of course (4a–d) seem wrong. But then it's hard to see how the phrase "verifiable under ideal circumstances of inquiry" goes beyond the simple claim that the rug is green.

What about "fits the facts"? The second take on (4a–d) holds here, it seems: it is hard to see how " 'the rug is green' fits the facts" goes beyond the claim that the rug is green – it is hard to see, that is, how further axioms for truth conveying the locution "fits the facts" would add constraints on the extension of the truth-predicate over and above disquotation.

The elusivity of truth is *epistemologically neutral* in pretty much the same way the disquotational role of the truth-idiom is; accepting Putnam modals, that is, does not support *any* epistemic view: realist, coherentist, pragmatist, and so on.

To see this, recall the distinction between cases where two terms are extensionally equivalent because of conceptual linkages ("vixen"; "female fox") and cases where we can *discover* two terms to have the same extension although they're not conceptually linked ("creature with a heart"; "creature with a kidney").

Putnam modals prevent *conceptual linkages* between "true" and non-disquotational notions such as "warrantedly assertible" or "verifiable under ideal circumstances of inquiry,"[19] but the modals, *by themselves, anyway*, do

---

19 Actually, Putnam modals show more than that the notions susceptible to them cannot be augmented or replaced: they show that the notions in question are (as I show in Part IV § 3) *criterion-transcendent*. But these forthcoming refinements do not affect the points being made here.

not prevent epistemic investigations from showing that (one or another) notion of inflated truth actually picks out exactly the same class of truths as "true" does. It might have been (might be!) that we were so lucky that our collective methods to pick out warrantedly assertible sentences picked out all and only the truths. Thus, although Putnam modals prevent a (partial) characterization *of* truth as what's verifiable under ideal limits of inquiry, they certainly don't prevent anyone from *claiming* that, in fact, every truth *is* verifiable under ideal limits of inquiry, or that every truth is whatever corresponds to the facts, or that every truth is whatever best coheres with our background beliefs, or even that every truth is whatever we as a community ultimately endorse. Putnam modals allow discoveries of this sort; but they do not allow us to tinker with the notion itself, and the question of whether what's true amounts to any of these things can be resolved, if it can be resolved, only by looking at our methods for *finding out* truths. It won't be resolved by an examination of the constraints on the notion of "true"; and all the elusivity of truth tells us is that such an epistemic discovery, if successful, will not contribute to a supplementation or definition of "true" (or a replacement for it) – unless, of course, we repudiate the *c* and *d* Putnam modals (in effect, dropping "true" for a homonym).[20]

Why are Putnam modals true? Here's how the question should be posed: what is it about our truth-gathering practices, our methods for finding out what's true, that makes it appealing to employ an elusive notion? Answer: we have methods for determining what we take to be true, and these are chosen for reasons of simplicity, maximal predictive value, tradition, and so on; but we may find any one method faulty in some respect, or badly applied; discover that the method either should be modified, or that the truth-values assigned to certain sentences on the basis of this method, should be revised. We may also add new methods, and adjust the truth-values assigned to certain sentences accordingly.[21] The elusivity of "true" handles these eventualities by allowing us to deny that a certain something, previously accepted as "true," is true, or that a certain something previously accepted as "false," is false.

Although our truth-seeking methods are no more than a certain collection of methods used and accepted at a particular time, what is true cannot be *defined* as what's establishable by these methods: we don't want to close off dropping these methods for other ones later, and, in contradiction of current usage, calling the results the later methods yield "truths."

That what we (currently) do to establish truths can't be offered as either necessary or sufficient conditions on what is true doesn't mean there's

---

20 We can't drop the *a* and *b* modals since they provide the adequacy test a disquotational predicate must pass.
21 After all, just these sorts of thing *have* happened.

nevertheless a non-disquotational characterization of "true" in logical space peculiarly beyond our reach. There's no such definition, and none needed: the truth-idiom is *supposed* to be open-ended as far as our truth-seeking practices are concerned. This property of the predicate "true" allows us to ask of *any* methods we have if they *really* are truth-seeking methods. On the other hand, at any particular time, there's nothing more to say about what's true that goes beyond a characterization of our current truth-seeking methods, because there's nothing more to truth, as far as we can see, than what we currently do. When new methods come along, we may adopt them. But we'll then be in the same position regarding the truth-idiom as we were before. It won't be that our new methods have somehow given us a better idea of what's involved (non-disquotationally) with "true."

Our practices with "true" differ from our practices towards what's currently warrantedly assertable *only* in that from a future vantage-point we *may* be able to say that although A is not true (and never was true – we only thought it was), still, it *was* a warranted assertion at the time we asserted it.[22]

The naturalistic fallacy suggested to Moore that "good," and other ethical terms, picked out non-natural qualities. (We have come to regard this view as misguided.) In the same way, upon discovering that "true" is not synonymous with any nondisquotational characterization, no matter how idealized, one may regard the term as picking out, non-naturalistically, some collection of sentences that can forever elude our grasp. This is a mistake (and if realism makes *this* mistake, so much for realism). The point to stress is that any such view is idle in two respects: it does nothing towards providing an explanation for our belief in Putnam modals; nor does it offer a description of what we do to find truths. It provides no insight about truth.

On the other hand, nothing about truth supports any metaphysical view, one way or the other, about whether representational views of reality are true or should be replaced with pragmatism of one sort or another, or whether realism–anti-realism debates are coherent, etc.[23] A disquotational notion has nothing to say about any of this – and a disquotational notion which is prevented (by Putnam modals) from being characterized in a more "substantial" way still has nothing to say about any of this. To decide what sort of grip, if any, we have on the world around us, and in what sense that grip transcends our own conceptual scheme (for, after all, why couldn't it, in the sense that any other conceptual scheme would have to grip the world in the same way?) is something to be decided by an analysis of what we do *epistemically*.

---

22 Whether this means that, "[a]lthough coincident in normative force ... 'T' and 'is warrantedly assertible' *have* to be regarded as registering distinct norms" (Wright 1992: 24), I leave to the reader.
23 See, e.g. Putnam (1978a, 1978b); Rorty (1986, 1993).

So let's return to "know." Having described the elusivity of truth, it's easy to see that there's nothing philosophically rich in the normativity of "know" either. For the normative flavor of "know" is due only to its conceptual link with "true," and the fact that what we currently do to gather truth provides neither sufficient nor necessary conditions on "true"; the fact–value distinction, at least in epistemology, is a prohibition only against conceptually linking "truth-laden" epistemic terms with terms not similarly "truth-laden."

What, then, does this imply, if anything, about the normative question regarding the methods we *should* use in epistemology? Well, first, the methods we *should* use cannot be conceptually linked to or defined in terms of the methods we *do* use; and, second, there's nothing more to our understanding of epistemology than the methods we *do* use. This explains why evaluative comparisons between epistemic methods must be made in terms which do not look intrinsically epistemic (e.g. degree of coherence, explanatory unity, survival, and all that); while at the same time we recognize that any of these modes of comparison can be one that we drop tomorrow because we subsequently deny it is truth-seeking.

It's not uncommon to criticize naturalized epistemological views on the grounds that they substitute for the genuine epistemic notion of evidence an unacceptable causal replacement. I illustrate this method of attack against my own approach. Consider ORC. There seem to be two possible interpretations of it. The first is to see it as a (rather uninteresting and uncontroversial) claim that theory construction must be *causally* constrained by evidence accessible to humans given their biological make-up (and this is why ob-similar sorts play the role they do). In one sense of "causal," this is not the right interpretation of ORC: for the implicit story about the human sensory system being restricted the way it is *itself* is a claim justified by scientific theory (in particular, our understanding of how our sensory system itself interacts with the natural kinds we currently find ourselves committed to). ORC starts further back, with ordinary epistemic practices regarding perception. I offer you a nice theory. You say: But you haven't explained *that*, where *that* is perceptually available to both of us. I'm then obliged by our ordinary epistemic practices to say one of the following: Oh *that*. *That's* (a) one of those group hallucinations we usually see; (b) something, given my current tools, which is too hard to explain *now*; (c) something that involves *other* things my theory isn't concerned with; or (d) (if Kuhn is right) *what* are *you* talking about? (*I* don't see anything). ORC relies on arguments that rule out (d); *each* of (a)–(c) places obligations on us – even (a) – we can't insouciantly invoke group hallucinations. These things, too, have ways of being recognized; perhaps it *isn't* one of those group hallucinations.

The reader intent on pressing the implicit normative idea of evidence might protest that I've described only *what we do*, not what we *should do*. That's right. (And here I raise again the point of the earlier part of this section.) There's nothing more to say now about what we should do except what we *do* do (according to norms we all have public access to). That might change later, but nothing more can be said about that now.

I've been concerned with one aspect of traditional epistemology that naturalized epistemology has been accused of overlooking: the *normative*. There is another aspect of traditional epistemology that some philosophers feel sits uncomfortably with naturalized epistemology – *scepticism*.[24]

First things first. Isn't "revision can strike anywhere" already scepticism? Quine writes (1981b: 180) about how we can know if one theory is true and another false:

> There is an obstacle ... in the verb "know." Must it imply certainty, infallibility? Then the answer is that we cannot [know]. But if we ask rather how we are better warranted in believing one theory than another, our question is a substantial one. A full answer would be a full theory of observational evidence and scientific method.

It is not clear why this answer should silence the sceptic. A successful sceptical argument, it seems, would undermine our beliefs in what we are better warranted in believing as much as it would our beliefs about the world itself. But Quine (and, consequently, naturalized epistemology) is better off with regard to scepticism than this reply suggests.

Quine describes scepticism this way (1975: 67–8):

> Rudimentary physical science, that is, common sense about bodies, is ... needed as a springboard for scepticism. It contributes the needed notion of a distinction between reality and illusion.... It also discerns regularities of bodily behaviour which are indispensable to that distinction. The sceptic's example of the seemingly bent stick owes its force to our knowledge that sticks do not bend by immersion; and his examples of mirages, after-images, dreams, and the rest are similarly parasitic upon positive science, however primitive.[25]

He continues:

> I am not accusing the sceptic of begging the question. He is quite within his rights in assuming science in order to refute science; this,

---

24 See Quine (1975, 1981a); Stroud (1981).
25 Quine's holism assimilates traditional scepticism to a scepticism originating in (rudimentary) physical science. Two-tiered coherentism, by allowing observational practices to survive the demise of science, must be undercut by a more traditional (Cartesian) scepticism directed towards perceptual procedures themselves, and what they yield. In what follows I waive these considerations.

if carried out, would be a straightforward argument by *reductio ad absurdum.*

Given this picture, Quine (1981a: 475) can claim the sceptic is overreacting: the sceptic assumes knowledge to refute knowledge by a straightforward *reductio ad absurdum.* Success in this endeavor, on Quine's view at least, is not particularly grave; at least no more so than the presence of paradox in set theory. There, recall, one's favorite collection of naive intuitions leads to trouble. Intuition now bankrupt, we turn to one *ad hoc* approach, then another, until the best solution is hit upon (where "best" is understood according to the needs of the mathematical community at large). Once a best solution is adopted, intellectual habit makes the next generation's intuitions straightforwardly favor this choice as a natural one.[26] In the case of a less formalized science, like epistemology, one's favorite intuitions about knowledge again lead to trouble. And again, intuition being bankrupt, one tries one description of our epistemic practices, now another, and, again, chooses the best possible, where "best" is understood in terms of what we need our epistemic idioms for.

This explains Quine's cavalier attitude towards "know": if to "know" something requires infallibility, but our best epistemic theory finds no such infallible beliefs, one mints another word to substitute for "know," and uses *that word* henceforth. Thus, naive epistemology gets supplanted by regimented epistemology, much as naive set theory is supplanted by formal set theory.[27]

This will outrage the sceptic if she feels the traditional subject is being avoided rather than refuted: if "know" is intuitively linked with infallibility, one wants to know *why*; only then can we be sure there's no (philosophical) danger in repudiating the link.[28]

I meet this challenge by attributing the source of the intuition that knowledge requires infallible methods to the sceptical tradition itself. This intuition is elicited by imagined situations where identically-applied truth-seeking methods yield false answers in one case, and true ones in the other. The intuition also relies on subtle mistakes in our appreciation of the epistemic qualities of the imagined situations.[29]

---

26 See Quine (1969d).

27 Oddly enough, I've found this sentiment towards "know" explicit only in Quine (1987: 109): "I think that for scientific or philosophical purposes the best we can do is give up the notion of knowledge as a bad job and make do rather with its separate ingredients."

28 Consider intuitions about the *a priori* status of mathematical truths. Even Quine doesn't dare to deny these *intuitions*; he subdues them by locating their source in the centrality of such truths to our web of belief. Should the intuition that a method yields knowledge only if it's an infallible method be treated with less respect?

29 This issue is also of importance because it motivates the inductive argument – that since every previous scientific theory has been wrong, it is overwhelmingly likely that our current

Start with "the argument from error." Pick a method you use for finding truths. This method sometimes gets you in trouble (hopefully you've noticed); and it does this because you expect one thing and get another. So: if the method sometimes gets you in trouble, without your realizing that it's getting you in trouble, then you don't know, any time you use it, that it isn't getting you in trouble *then*.

This argument has been pretty thoroughly run through, and there are many moves philosophers have used to short-circuit it; moves, by the way, that are usually pretty implausible. But notice that this argument isn't convincing at all when applied to methods one-by-one. Two examples: our senses sometimes fool us (railroad tracks seem to converge, equal lines seem different in length, etc.), so use of our senses seems to be a fallible method in the above sense. Also, when calculating sums, we (most of us) often make small errors and get wrong answers. Calculation, too, seems to be fallible in the above sense.

These considerations don't convince anyone of scepticism – not by themselves, anyway. Why not? There are two reasons. The first is that we recognize that we apply epistemic methods together, not discretely: they supplement and complement one another (especially with regard to their individual failings), and so the results of their piecemeal application have cumulative power. I may check an addition sum, derived one way, with other methods of computation, friends, a trusty computer, etc. Similarly, I supplement visual impressions by trying other angles, bringing other senses to bear, etc.

Before turning to the second point, I should dwell a moment on the particular sort of coherentism at work here. Coherentist methods are often seen as a matter of our web of belief hanging together in a logically consistent way. But, equally important, especially for an appreciation of how the lower tier of perceptual regularities and perceptual procedures coheres, is that these things have restricted domains of application which complement each other. For example, I may not be sure of the properties of something visually – and I use other senses to cement my views of the object. It's not so much that these procedures correct each other as that they have penumbra where they are more or less to be trusted, more or less in need of something else to supplement them. Observational regularities too, are linked together in that we recognize, when we're at the edge of the scope of one of them, that another should be brought in, and which. Part of our tacit grip on observational regularities is recognition of when and how they go together in applications.

scientific theories, too, are wrong – used to support "approximate" views of science such as Boyd (1981), Giere (1985), and, much earlier, Popper.

136

The second point is that we not only correct our methods when they fail: we often *explain how* failure came about to begin with. Mistakes, that is, are not epistemic black boxes that threaten everything we know, but are localized by recognition of how they arise. This is understood, both in the sense that an explanation for them is given, and in the sense that the pattern of their appearances can be identified.[30]

If this is right, then the sceptic can press her case successfully only by undermining *all* our epistemic methods *at once;* this explains the popularity of demons and mad scientists.[31]

Contrary to what the history of epistemology suggests, we're actually *instinctive* coherentists. We're also fallibilists. We're prone to say, when first faced with sceptical arguments about misleading senses or computational mistakes: "Sure, anyone can make a mistake about anything. So what? Mistakes can be corrected, and *we do it all the time.*"

There are two ways to take demon hypotheses. The first, and weaker, way is to take such possibilities as showing that any particular knowledge-claim might be faulty because it's possible for the demon to fool us about the claim (imagine a situation where nearly everything is as we take it to be, except that the demon intervenes now and again, arbitrarily, to throw us off). This version is easy to respond to: we *always* make mistakes (anyway), and we know how to correct for them. The second, the stronger, way is to take demon intervention *globally*: we could be wrong about everything, nearly

---

30 Even if we don't understand how certain optical illusions operate, we may still recognize when they're likely to occur. The mathematical example is similar. I may know my weakness for certain computational errors, and learn to anticipate them.

An illustration of a treatment of mistakes as a "black-box" phenomenon, something which we cannot learn to anticipate and test for, may be found in Collins (1985), where he talks of "open systems" in describing, for example, how experimenters fine-tune their mastery of lab apparatus. Actually, seasoned experimenters know that not anything can go wrong, and when we've learned a lot about a particular situation, what we've learned is how, typically, things do go wrong. This is part of the learning curve involved in mastering experimentation techniques of any sort, and this mastery is not a matter of being socialized the right way, but learning how (this bit of) the world operates.

By the way, the demand, expressed earlier, to explain errant intuitions, such as *aprioricity* or the requirement of infallibility for methods that yield knowledge, is itself an illustration of a practice for diagnosing mistakes, a practice the sceptic, and those who ply his tools, overlook. Even the set-theoretical intuitions giving rise to set-theoretical paradoxes are commonly supplied with an explanation: hasty generalization.

31 This makes Descartes (1641) seem disingenuous. He raises, first, the weaker sorts of consideration, about how our senses sometimes mislead us, and makes it look as if some purported knowledge is in doubt, though not all such; and so he passes on to the dreaming argument. But I'm suggesting that unless he undermines our knowledge claims *all at once,* his considerations about our senses have no sceptical force at all. Demon arguments and their ilk are needed not for epistemic full coverage but to get the sceptical project going to begin with.

enough. It's only if the demon possibility is interpreted *this way* that infallibilist intuitions about knowledge emerge. For only with infallible truth-seeking methods can we escape the epistemic catastrophe so poised.

But doesn't this last observation show that our knowledge practices *do* require infallibility (and only when we meditate along with Descartes, as he invites us to, do we see this)? Not necessarily. What it may show is that we require our truth-seeking methods to be correctable (in principle). The problem is that, when we consider a demon scenario, the only truth-seeking methods which seem correctable there are ones that don't cause trouble in the first place, i.e. infallible ones.

It might seem that I've won the battle over "know," but lost the war over knowledge. For the sceptic can simply give me the word "know," and argue that naturalized epistemology, put my way, concedes the sceptic's argument: the natural epistemologist must agree with the traditional sceptic that infallibility is out of reach. I decline to take this route, however. Sceptical scenarios, as traditionally (naively) posed by the sceptic, are not quite acceptable to the naturalized epistemologist; and the sceptic cannot quite draw from them the lessons that he wants. I show this in Part IV § 5.

Part III

PERMUTING REFERENCE

I see no gainsaying the proxy functions.

(W.V. Quine)

§ 1

INTRODUCTION TO PART III

Recall the program–scope gap from Part I. I described it as a split between what we *say* (our scientific theories and what they range over) and what we *do* (successful application of theory). This split also lives among the rarified idioms of truth and reference. One way to see why this would be is to recognize that these idioms are part of an (applied) *theory* of language that has been mathematized in recent years (analogously to how folk physics was mathematized centuries ago). The program–scope gap arises in *disguise*, though: as philosophical puzzles about naturalizing reference.

I'm sympathetic toward attempts to "naturalize" reference: attempts to explain how humans, given their limitations, can nevertheless refer to what they do refer to.[1] Such explanations are supposed to reveal the mechanisms underwriting reference.[2] Sadly, as Part IV shows, such attempts must fail at full coverage, at least in so far as ordinary semantic assumptions are concerned: (most of) our terms have many referents that are without available appropriate mechanisms to reach them. Our referential practices, that is, seem to presuppose a capacity to refer to items that *no* naturalistically acceptable (i.e. broadly causal) explanation can handle.

The next best thing for the causal theorist, given this glum conclusion, is to provide as much of a causally acceptable explanation for our referential

---

1   We have paradigmatic examples of such "naturalistic" explanations. My use of "Einstein" refers to Einstein because of the right sort of causal chain between my use of the word and Einstein himself, a chain starting with my learning the word from others in the right way, and traveling via similar acts of learning to a "baptismal ceremony" where the word was first applied to Einstein. This idealized picture fits a large number of cases, if "baptismal" ceremony is taken broadly, and if other items are admitted into causal chains: written materials, photographs, video tape, etc. Similar paradigmatic "naturalistic" examples for natural-kind terms, and what they refer to, are available, which involve acts of generalizing the right way from "baptisms" of samples.

2   Here, and in what follows, when I mention the causal theory of reference, I mean a family of positions, including those on which reference cannot be fixed by causality alone but must involve descriptions also. See e.g. Boyd (1980) or Lewis (1984).

practices as possible: we should be able to measure how far our referential powers (causally construed) do extend, and how far we're just bluffing. After this investigation, we can then evaluate various forms of *revisionism*: proposals to modify our ontology and/or theory of truth to bring the referential apparatus of our language practices within naturalistically acceptable bounds (as construed by the causal theorist); or we can conserve, and give reasons why our talk of reference and truth, despite the presence of a certain amount of sheer bluff, should be left alone.[3]

The next best thing is *not* possible if we *cannot* distinguish those "made-up" or "free-floating" aspects of our referential talk from those aspects supported by naturalistically acceptable mechanisms. Some philosophers believe that this distinction *cannot* be made out – Quine, the later Putnam, and Davidson, among others. I think this view is very wrong; but it's beyond me to go through *all* the arguments marshaled for these claims and rebut them. I'll do something more bite-size: trace the history, and evaluate the quality, of a particular family of arguments often used to obliterate the distinction. This family may be loosely described as "permutation" argumentation: arguments of this family try to show that any referential scheme we might give for our language is open to a certain kind of *systematic* reinterpretation or permutation. On this view, the particular items our terms are purported to refer to, in some sense, exert no force on the form our language takes: *reference is inscrutable*.

Permutation arguments are very popular and come up again and again.[4] The focus on them (including versions using non-standard models generated via the Löwenheim–Skolem theorem) has obscured what's really motivating philosophers who cite one or another permutation argument as the official reason for a position held. Apart from indicating what's wrong in these arguments, I set out the conceptual geography to make clear what depends on what.

After discussing permutation arguments, I raise a quandary for the causal theorist. This is that it seems clear we could *explicitly* adopt referential schemes that violate naturalistic causal constraints on reference. The point is *not* one raised by the permutation theorist, as I'll show: It's *not* that reference is inscrutable. Rather, it's that it seems open to us to officially shift what we refer to in ways that shouldn't be admissible if the causal theorist is right.

---

3 Revisionist programs can involve changing our logic: replacing sharp two-valued predicates with vague ones. Mathematical nominalism (on the grounds that there are *no* naturalistically acceptable mechanisms which explain how we refer to mathematical objects) is another example of a revisionist program. Arguments for taking a conservative attitude towards our empirical talk are given in Part IV.
4 See the list in note 10 to the General Introduction.

Outline of Part III

In § 2, I briefly review those properties of formal languages that give birth to permutation arguments, and lay out general methodological points. In § 3, I turn to the classical discussion of permutation arguments, in Quine's work. In §§ 4–5, I cover the later versions of these arguments by Field and Putnam. In § 6, I discuss the ontological status of the sort of causality that causal theorists like, and defend it; in § 7 I raise puzzle cases, and in § 8 I take stock.

§2

FORMAL CONSIDERATIONS

Logical preliminaries

Any countable, consistent, first-order theory has a countable model. Moreover, given any uncountable model of such a theory, a countable model may be extracted from it. This is one version of the Löwenheim–Skolem theorem. It's sensitive to the resources of the underlying logic, vanishing if we go second-order, modify quantificational power, or alter the upper cardinal boundary on sentence length. *But* it's a deep result for all that, since such modifications can be costly.[1]

Given any countable, consistent, first-order theory and a model of that theory, consider *any* permutation of the domain of the model onto itself. By similarly permuting the referential relations of the names and predicates of the language, we can construct another model of the same theory. This property *is* humdrum: it generalizes to second-order logic, and the other logics mentioned, though the Löwenheim–Skolem theorem doesn't.[2]

Some technicalities

Given an arbitrary permutation $f(x)$ of the domain of the model onto itself (or a 1–1 mapping of the domain of the model M onto another set of the same cardinality), a new model, M', is defined thus: if the name **c** is mapped

---

1   A precise statement of the theorem, its proof, and interesting strengthenings may be found in Chang and Keisler (1973). They call it the Löwenheim–Skolem–Tarski theorem for obvious reasons. I retain the shorter nomenclature because that's how the theorem is commonly referred to by philosophers. One or another version of the result can also be found in most logic textbooks, e.g. Boolos and Jeffrey (1989); Enderton (1972); or Shoenfield (1967). See Barwise and Feferman (1985) and van Benthem and Doets (1983) for details on those logics where versions of this theorem vanish.

2   It's easy to design logics where this otherwise trivial theorem *doesn't* hold. See Azzouni (1991). But this is an unusual case where there's a two-place device present with truth-conditions that include the constant appearing in its first place is mapped to the sentence appearing in its second place.

by M to the object c, then M' maps c to f(c). Similarly, if M maps an n-place predicate P^n to a set of n-tuples $(p_1...p_n)$, then M' maps P^n to the set of n-tuples

$$(f(p_1), ... , f(p_n)).^3$$

Each sentence of the language is true in M iff it's true in M'.

We may consider certain generalizations. The above 1–1 mappings always guarantee that the desired sort of model can be defined on the target set. But in certain cases the mappings need not be 1–1 and a model can still be forthcoming.[4]

For historical reasons, call the permutation function a "proxy function," and call the minor theorem just sketched "the humdrum theorem." Also, call any mapping determined by the Löwenheim–Skolem theorem that takes the domain of M to a countable subdomain of M a "proxy function." Proxy functions due to the Löwenheim–Skolem theorem are identity mappings which fail to be defined on all objects in the domain of M (some of the domain, that is, is ignored by the function).

What philosophical implications do proxy functions have? Start with the Löwenheim–Skolem theorem case: if we have no more resources for linking our terms to the world than those supplied by asserting truths in countable first-order languages then (for all we know) our predicates hold at most of countable numbers of things even if we think they don't; and a model for our truths can be defined on *any* countable collection of items. This may be regarded as a version of Pythagoreanism.

The antecedent of this conditional is dubious: *why* do we have no more resources for linking our terms to the world than those afforded by first-order languages? Why couldn't it be, for example, that we have second-order resources?[5] An argument based on the humdrum theorem cannot be escaped this easily. Nevertheless, *prima facie*, such arguments seem to bear, if at all, only on our capacity to refer to *mathematical objects*, but to have no effect whatsoever on our capacity to refer to empirical objects. This is because it is natural to assume that we have additional resources for referring to empirical objects: the causal connections forged and exploited between ourselves and such objects – while such connections are patently absent in the mathematical case. The expressive resources afforded by a logic are purely formal:

---

3   If the language contains function symbols, a similar definition is available for them also.
4   For some examples in the first-order case, see Chang and Keisler (1973). Also see Quine (1969: 56–7).
5   Quine (1964, 1969c) considers various strategies to avoid Pythagoreanism, but never considers going second-order. Putnam (1977) relies on the Löwenheim–Skolem theorem, which restricts him to first-order logic, but later (1978b, 1981) uses an argument based on the humdrum theorem.

anything fitting the conditions of a set of truths will do. But causal relations are *material*: they *nail* our usage to *certain* particulars, even if *other* things have the same formal properties.

These arguments really do seem to have clout against *abstracta*, provided *abstracta* are construed in typically Platonist fashion. But it takes, or *should take*, auxiliary argument to show that we have no additional (causal) resources to help fix the references of our terms to natural kinds and concrete individuals. *And* the general form of such an argument should look like this: proxy functions provide admissible alternative interpretations for our natural-kind terms and proper names, which are compatible with all our resources for fixing reference because ..., where what follows "because" are considerations ruling out other possible resources for fixing the references of our terms. Instead of this acceptable argument-form, or so it seems, the mere *existence* of proxy functions (in one guise or other) is often offered as an *argument for* the claim that there are multiple interpretations possible of such terms, and, almost as an afterthought, considerations are (sometimes) added to infirm other possible resources.

What could motivate philosophers to regard proxy functions *all alone* as convincing? Here are two possible types of error:[6]

1   Suppose a (causal) *mechanism* is taken to fix reference. Its capacity for this *can't* be denied on the grounds that the *language* describing the mechanism may be reinterpreted by a proxy function. If I claim our terms are fixed by mechanism M, the response that the term 'M' can be reinterpreted to mean something else misses the point. It is not the *term* 'M,' or even the theory 'M' is embedded in, that fixes reference on this story: it is the *mechanism* referred to by 'M' that fixes reference. For historical reasons, call this *Devitt's complaint*.[7]

2   Suppose the possibility of exchanging "dog" and "cat" throughout a language is raised. *Without further argument*, this possibility shows

---

6   Convention for what follows: "M" is a variable ranging over mechanisms; "'M'" is a variable ranging over the *names* of such mechanisms.

7   Glymour (1982) and Devitt (1983) independently direct this complaint at Putnam (1977 and 1978b). In calling it "Devitt's complaint," I'm not suggesting that Devitt was the first to point out the error or the first to attribute it to Putnam. But he dwells on it more than do others as a diagnosis of Putnam's penchant for permutation arguments. Glymour explicitly describes the purported error as a use/mention one. Also see Devitt (1991: Chapter 12), where he repeats his charges. The particular mechanism postulated *could be* one that fixes the reference of the term 'M' by way of the theory in which 'M' is embedded. In this case, provided one had already shown that the resources of the theory for fixing the references of its terms were insufficient to rule out alternative interpretations afforded by proxy functions, one could conclude that 'M,' too, is open to reinterpretation. But this wouldn't be how to *start*.

nothing peculiar about how we pick out the animals we currently call "cats" and "dogs." Consider a standard three-dimensional Cartesian coordinate grid for the entire universe. The origin can be placed anywhere, and each location chosen changes the coordinates of every other location. Nevertheless, *without further argument*, this shows nothing special about the world or about our capacity to refer to items in the world via this coordinate system. For historical reasons, call this *Alice's rejoinder*.[8]

These errors are related, though for purposes of textual analysis it can be important to distinguish them. If a case can be made that an argument is open to Devitt's Complaint, however, the case can usually be modified to show that it is open to Alice's Rejoinder too.

These errors should not be confused with legitimate attempts to show mechanisms associated with "reference" do not fix the references of our terms, of course. But the methodological point is simple: One must *establish* grounds that these mechanisms fail *before* one is licensed to take referential schemes to be permutable. The neat clarity that permutation arguments have in the writings of their proponents often turns on rhetorically reversing the proper order of argument.

---

8 "When *I* use a word," Humpty Dumpty said, in rather a scornful tone, "it means just what I choose it to mean – neither more nor less."

"The question is," said Alice, "whether you *can* make words mean so many different things."

(Carroll 1871: 186)

§ 3

QUINE'S VERSION

I start with Quine's essay "Ontological Relativity" (1969c), where he first applied proxy functions to a non-mathematical context. Quine begins by reviewing the "gavagai" example from his *Word and Object* (1960). A rabbit hops by, a native exclaims "Gavagai," and we wonder what's being referred to.[1] The famous problem is "that a whole rabbit is present when and only when an undetached part of a rabbit is present; also when and only when a temporal stage of a rabbit is present" (1969c: 30). As a result, "we can never settle the matter simply by ostension ... simply by repeatedly querying the expression 'gavagai' for the native's assent or dissent in the presence of assorted stimulations" (ibid.: 30–1). Why not? Because "[t]he only difference is in how you slice it. And how to slice it is what ostension or simple conditioning, however persistently repeated, cannot teach" (ibid.: 32).

It's worth stressing that up until this point, causal connections between word-usage and the world *have been* Quine's focus. Factors such as salience, "one's own inborn propensity to find one stimulation qualitatively more akin to a second stimulation than to a third" (1969c: 31), have not been excluded from our tools to individuate what natives talk about: it's just that Quine thinks such resources won't do the job. Consequently, he despairs of looking to the native's interaction with the world to fix the reference of the native vis-à-vis the various rabbity options. All that's left to decide the issue is the projection of certain *grammatical forms* onto native utterances: the individuation of native words must be handled by *our* quantificational devices, and a residue of arbitrary choice in the imposition of such devices on native speech remains.

Those planning to couch the referential attributions they make to natives via the causal connections they see natives make to the world should be justly puzzled. How did we get here so fast? One popular diagnosis blames

---

1   I separate, on the causal theorist's behalf, issues about what the native is referring to from questions of how to translate "gavagai." Translation involves concerns beyond those raised by reference, as I explain later in this section.

Quine's linguistic behaviorism: he thinks we only have access to *surfaces*: verbal and non-verbal behavior against backgrounds. This exegetical gloss is refuted by Quine's explicit acknowledgment of psychological dispositions. No, what is going on is rather that Quine has neatly excluded the causal relations themselves that the causal theorist hoped to rely on to fix reference. Given Quine's set-up of the situation of radical translation, such relations are not perusable.

The crucial move is not explicit in "Ontological Relativity" (1969c); one must travel back to Chapter 2 of *Word and Object* (1960). Quine, good physicalist that he is, thinks that during radical translation we must focus on the sensory stimulations of the natives: the excitement that external events cause in the natives' nerve-endings. This requires an antecedent *modulus of stimulation*: a decision on the appropriate lengths of the time-slices of nerve-excitement which are to demarcate relevant causal (and referential) factors when natives utter sounds.

Precisely this move, however, unjustifiably deprives the causal theorist of what he needs. Talk of sensory stimulations, physicalistically acceptable though it may be,[2] is talk neither field linguists (nor anyone else for that matter) have access to. One can speak *in general*, as Quine does, of sensory stimulations. But without an explicit vocabulary in terms of such stimulations, *the scope of which extends easily to natives*, this is useless. Such a vocabulary would require labels for each nerve-ending (or significant group of nerve-endings); and the linguist would need methods (instruments) to tell in each circumstance *which* nerve-endings were "on" or "off." Nothing like this is available.[3]

What tools *does* the field linguist have? Perceptual procedures, with a few additional instrumental aids, such as recording devices, video cameras, etc.; theoretically, whatever from the human sciences (biology, linguistics, etc.) can be brought to bear. That is, as in the stereotypical picture of radical translation, she faces (or is within hearing of) natives, jungle plants, rabbits, lions, etc., all of which she can refer to in enough detail to mark out regularities in the native's experiences which are simultaneously observational regularities for her. Given her rudimentary technology (from the point of view of an experimental physicist), her vocabulary is largely restricted to core terms, and what she can study is restricted to cores: the stimulations

---

2   Talk of sensory stimulations belongs to the special science of physiology. I waive the issue of how such talk can be reduced to physics. Notice, though, that the program–scope problems raised momentarily about this kind of talk apply even more to nerve-endings and pristine physics.

3   Davidson (1973: 136 note 16) repudiates Quine's stimulations for "reference to the objective features of the world which alter in conjunction with changes in attitude towards the truth of sentences ..."

relevant to what the native is referring to, therefore, *must be* expressible in these terms. Nerve-endings don't come into it.[4]

Let's step back a moment to see what the causal theorist's approach requires, and whether the resources I've described suffice. There are three assumptions.

(1) Referential relations are constituted out of causal relations.
(2) There exists a rich enough set of causal relations to give *individuation conditions* for the causal *relata* of native utterances; they distinguish among candidate items, such as undetached rabbit parts, rabbit, rabbit fusion, etc.
(3) These rich causal relations are epistemically accessible to the field linguist, given the tools (and the science) at her disposal.

All *three* assumptions can (and have) been attacked. I'll say a bit about objections to assumptions (1) and (3), but it's important to realize that Quine objects only to (2), and his objection – in pretty much his version – continues to attract philosophers, even those otherwise alien to his viewpoint.[5]

Assumption (1) can be attacked by arguing that reference cannot be constituted out of causal relations because referential relations have normative properties that causal relations do not. Such a gap between causality and reference can be argued for in several ways: one is to claim that causal relations aren't sensitive to mistakes about what we're causally interacting with, although referential relations are – we can have causal relations to rabbits even though *we think* we're trafficking with rabbit-fusion. In such a case we refer to rabbit-fusion even though the causal theorist – plying causes to construct reference – would conclude we refer to rabbits. (This objection accepts assumption (2) for the sake of argument, and goes after assumption (1).)[6]

---

4   What if the native is a (non-human) *alien* and is biologically sensitive to things the linguist is not? Then, without instruments that enable her to recognize regularities outside her sensory range but within that of the alien, she will fail (to some extent): the alien will sometimes seem to be referring to nothing at all.

5   Richard writes:

> Some would appeal to causal considerations to defeat inscrutability: Reference, we are told, is grounded in causal contacts between applications of word and object referred to. It is not clear that this is much help. Touch a cat and you touch an undetached cat part. Smell a dog and you are causally in touch with the dog fusion.... Once this occurs to one, it becomes somewhat plausible that there may not be *any* physical fact that distinguishes one of two incompatible reference schemes as *the* correct scheme. If we think that the physical facts determine all the facts, we will then find inscrutability a plausible if unsettling doctrine.
>
> (1997: 166)

6   I take this argument very seriously, and explore its impact on causal programs in Part IV.

Related to this attack is what we might call a *Davidsonian strategy*. Causes are part of our scientific world-picture. Reference, by contrast, must be projected onto natives via a charitable interface with a simultaneous attribution of beliefs to them. Consequently – even if assumption 2 is right – reference cannot be *constituted* out of causal relations.[7]

I am not going to offer much of a defense of the causal theorist's commitment to assumption 1 – especially since I'll undermine it myself in Part IV. But I should say now that I emphatically do not agree with the Davidsonian strategy; I do not think, that is, that belief and reference intertwine in a way that threatens assumption 1. Unfortunately this is a topic I cannot take up now,[8] though I will say that the causal theorist can undermine the Davidsonian strategy by attacking the Davidsonian assumption that reference is a theoretical relation (see § 6).

Still, I want to say *something* more about how the causal theorist should defend assumption 1. He should borrow a page from Part II § 7, and argue that the normative flavor we sense in *reference* is there because of its links to truth. Thus he should argue that there is *nothing more* to reference (as with truth) than what we *do* to refer (and this must be cashed-out causally) although, as with truth, "refers" is elusive for the same pragmatic reasons that make elusivity of "true" appealing.

Assumption (3) can be attacked by a variant of the Davidsonian strategy against (1). To recognize *how* natives are affected causally, we must attribute psychological states to them. But these, or many of them, are *beliefs*; and so access to the very causes that the causal theorist needs are themselves inextricably tied up with belief-attribution. Thus these causes are not robust items independent of variations in how we attribute beliefs to natives, but the softer result of the holism of interpretation.

---

7   This must be distinguished from the first attack against assumption (1), because Davidson's commitment to charity prevents him pressing considerations that turn on natives getting things (drastically) wrong. Davidson argues that the notion of reference plays a theoretical role *entirely* restricted to the application of a truth theory. Thus it's a theoretical notion to which we have neither observational access nor access through other theories (apart from truth-theories). This makes reference quite atypical: access to robust theoretical entities is not normally trapped only within one theoretical framework. See e.g. Ellis (1985: 58 note 6). I think this is a symptom that Davidson has misconstrued how attributions of reference (and belief) are made.

8   I'll cram some remarks into this footnote, and hope for a later opportunity to spell them out with the care and detail they obviously deserve. When two sorts of thing contribute causally to a phenomenon, two kinds of particle to a force on something, say, they can often be sorted out by defeasibility conditions of a certain sort: if we tweak things this way, we get *this* effect, otherwise *that*. Davidsonians too quickly impose charity – of one form or another – as the sole tool for analyzing how the interplay of reference and belief leads to "honest utterances." Ordinary speakers are actually quite sharp about distinguishing the causal effects of public language (where reference lives) from the idiosyncratic effects of psychology (where belief lives). I apologize for the brevity, and perhaps the cryptic quality, of these remarks.

Another and related way to attack assumption (3) is to grant epistemic access to the causal relations constituting reference, but deny that the particular causal relations *actually* constituting reference can be distinguished from other causal relations *without* attributing beliefs. After all, there is much discrimination (selective response to causal imputs) that doesn't emerge into the light of reference. (We can react to things we aren't quite conscious of, and we need not even be *able* to refer to these things.)

The response to both attacks is that the Davidsonian has the *evidential picture* backwards. Recognition of the appropriate causal factors *precedes* psychological attribution, and does not rely on it. The robustness of the pertinent causal factors and our access to them are *masked* by the language describing the relevant causes, because such language is invariably psychologically *loaded*. However: the apparently holistic latitude in psychological attribution encloses a rigid core of appropriate causal relations that underwrite successful psychological attributions.[9] Thus the causal theorist concedes the need to attribute beliefs, but denies this infirms the robustness of the causal factors recognized or our ability to get at them.

Let's return to assumption (2). What causal relations can *individuate* the object being causally interacted with in a way that the causal theorist needs? If we're restricted to perceptual relations, we'll not find *any*: these relations are, at best, to the surfaces of things. The causal relations needed are those at work during the *entire history* of the interaction of a speaker (and his community) with a kind of thing.

If I point at the ocean or at a rabbit, my causal relations at that moment are pretty limited in both cases; on the basis of these alone, there's nothing to reveal whether my interaction with the ocean is with *one big thing* or with *lots of little things*, and the same is true of rabbit fusion. Once my interactional capacities to make distinctions among things are brought to bear on the situation, and my causal history with regard to the ocean, and to rabbits, is perused, it can be seen that the ocean – causally speaking – is one big thing for me, and that rabbits are not.[10]

9 I fear this response to Davidson is cryptic, too. Two points: first, the allusion to my earlier discussion – of naked propositions expressed by sentences (of a conceptual scheme) that have additional content – is deliberate; second, a way to see how psychologically loaded language with wider content can cover a more rigid recognition of causally pertinent data is to notice that *behavioral facts* about our fellow humans are usually conveyed in psychologically loaded terms. We may accept that a particular psychological attribution we expressed is wrong (as a whole) and endeavor to convey exactly the *same* behavioural *content* with an otherwise different psychological attribution. Again, this is a matter I just can't pursue now.

10 If I had *really* fine-tuned senses – *and* really fine-tuned physical powers to act on the distinctions my senses could make – I would be able to causally interact with *water molecules*. In that case, my causal relations with the ocean would be similar to my relations with rabbits: my causal powers would individuate both rabbits and water molecules.

This should make clear that the causal relations beings have to things – if seen from a broad enough context – are rich enough to individuate things in quite different ways – and it's easy to see that creatures with different powers individuate things differently.[11]

The causal theorist is not committed to a metaphysical claim about there being a fact in nature about how things – the things anyone refers to – are individuated. He only has to say that – in most cases – the causal relations between a speaker (and her community) and a kind of thing provide individuation conditions: they tell us the causally relevant units the speaker interacts with. We can see that this should be true if only because the tools used to uncover these causal relations and what they pick out are primarily gross regularities.

Because of this the examples pertinent to Quine's discussion – rabbits, rabbit-stages, rabbit-fusion – are kinds with pretty central ob*-similar kinds as cores; and it is by using observational regularities about these cores that we learn *which* cores natives are in causal touch with.[12]

This leads naturally to assumption (3). Quine is a confirmation holist, so he recognizes – in principle – that data can come from anywhere. But his narrow view of the pertinent causal relations between native speakers and objects referred to makes him overlook how much information the field linguist has access to, information that (a) bears on the actual causal relations between speakers and objects, and (b) is from *outside* the speech situation. Evolutionary considerations provide constraints; anatomical studies do something similar: we know what bats are and are not sensitive to – causally speaking – in just this way.

Imagine the field linguist in search of data for the reference of "gavagai," and consider the two (empirical) candidates "rabbit" and "undetached rabbit-parts." How might the careful investigation of the relationship between natives and these rabbity things incline her towards rabbits as the appropriate causal *relata* of "gavagai" and not their undetached parts? One needs a feel for the texture of native life. Imagine, for example, that the interaction of the natives with rabbits is quite complex – they not only eat rabbits, but keep them as pets. They act as if rabbits have awareness, where this is understood in the ordinary way: the thing reacts, eats, notices what's going on around it, must be hunted, sometimes tricked, etc.[13] The native, in

---

11 Consider a fish being attacked by *Pfiesteria* – single-celled aquatic organisms (see Burkholder 1999). The fish interacts with *Pfiesteria*-fusion because it does not have the causal powers to distinguish single-celled items. We, too, until recently, were in the same boat as the fish whenever *It* attacked *us*. But we now have instrumental access to the separate organisms, and can individuate *Pfiesteria* differently.

12 On the causal story, causal relations between uses of terms and what they refer to always involve, strictly speaking, more than what's observationally accessible to speakers, as Part II made clear.

13 Watching someone hunt rabbits tells us volumes about the kind of thing she's interacting

this case, is clearly not merely (or, sometimes, even at all) concerned with the parts the rabbit carries along when it travels, but how the creature as a whole is operating.

How might we come to know this? *Not* by translating or interpreting their sentences, and not by imposing an apparatus of quantification on their utterances, but *rather* by getting a grip on what their actual (causal) interaction with rabbits amounts to.

What's been described thus far doesn't quite exclude "undetached rabbit-parts" because it's still possible that the causal *relatum* of the term is a complex kind that includes the entire rabbit itself among the undetached rabbit-parts; for what we can see of native interactions is satisfied either by assuming the term "gavagai" relates only to rabbits, or by assuming that the term actually relates in addition to all the further instances of undetached rabbit-parts that lie buried in each rabbit. Here a quite justifiable version of Occam's razor is called for. If a full description of the causal relations between someone and the world is possible in terms of a *simple* kind, then we should exclude more complicated but causally idle kinds.[14]

A similar argument can be used to rule out rabbit-stages, if natives are not causally interacting with them. However, the field linguist might conclude that rabbit-stages *are* the appropriate causal *relata*. This can happen because the native interactions in question are *minimal*. Suppose the natives see a particular kind of antelope *only* at dusk (because of the animal's shy and nocturnal habits), when it's particularly misty. And suppose they've a label for it. The natural first reaction is that they have a name for the *animal*. But that's to confuse *our* causal relations with the animal with *the natives'* causal relations. Their causal interactions with it, we might discover, are not like their causal interactions with animals but more like

with. Do we need a *theory* of natives here; a psychology of natives? Not in any scientifically respectable sense: we do need to be able to recognize when the hunter is sneaking up on the rabbit (why she thinks it's necessary to sneak up on the animal), and we need to be able to tell when the hunter thinks she's made a mistake – we need gross regularities (about natives). Does the success of this knowledge-gathering endeavor call for translation or interpretation of native sentences? Hardly – pretty much the same insights are needed to understand what dogs, cats, and snakes are up to when *they* hunt.

It's now palpable why Quine's talk of "moduli of stimulations" (1960) seems so artificial. Make the modulus short, and enormous amounts of causally-relevant historical information are lost; make the modulus long, and we're overwhelmed with *irrelevancies* that we have no means of picking through. In practice we're selective and picky about what we look for as candidates for what natives causally interact with, and we're guided in this by our grasp of observational regularities.

14 This principle must be applied with care, because even the causal theorist admits that speakers need not have causal connections with *all* the members of a kind. Eventually I show how this last observation causes trouble for the causal theorist – but the resolution of these troubles will not resurrect the Quinean view of these matters.

causal interactions with dreams or group hallucinations. In this case, we're likely to attribute to them the belief that an antelope isn't an animal but a kind of hallucination or a deity – perhaps their causal interactions with separate appearances of the same animal are not even bundled together by them – in this case sightings of *the same animal* would have to be individuated differently for them.

Insight into the causal interactions that determine native principles of individuation do not (usually) wait for translation of their idioms of identity into ours. Rather, we can often *just see* when natives think they're facing the same thing and when they don't. Also, we can tell a lovely evolutionary story for why their survival turned on recognizing identity in certain cases. Notice how our own perceptual procedures, and scientific theories, when relevant and when applicable, offer both strong constraints and an implicit understanding of what causal relations are involved.

Consider rabbit-fusion. We're forced to treat this as the appropriate causal *relatum* if native causal interactions with rabbits are so meager that rabbit-fusion *is* the *relatum*; this again is by the causal version of Occam's razor.

Here are some details. The difference between "rabbit-fusion" and "rabbit" is that the former is a mass noun, and the latter is not. This grammatical distinction translates (causally) into counting practices; that is, the distinction between the causal *relata* of count as opposed to mass nouns manifests as different ways in which our practices associated with these words impinge on the world.

Suppose natives hoard certain discrete food stuffs (such as apples), and that stealing sometimes takes place. Furthermore, suppose we see that natives often do something that *looks like* counting (they go through a process of picking up each item and putting it down in a different pile), and that they usually recognize when something is missing *even though* the number of items can be quite large.

That they are *counting* is an empirical thesis, of course, and other hypotheses must be ruled out. For example, perhaps they are actually measuring volume or mass. We can test these alternatives if the items are generally irregular in size or mass by making appropriate substitutions in their hoards and seeing if and when they notice differences. It is exactly the sorts of causal relation natives have to collections of foodstuffs that tell us what individuation conditions they're using; and it's nice to be able to say that this can be recognized for what it is without having any grasp of their language at all.

Our ability to grasp details of native practice (i.e. counting) enables us to determine, in certain cases, on the basis of the causal *relata*, whether the terms so causally related to the world are count or mass nouns; in particular, we have a principled way to exclude "rabbit-fusion" as a candidate for "gavagai," if, indeed, that is not what they causally interact with. And this procedure does not rely on translation – but precedes it, and may be used as data for translation.

"Gavagai," of course, is *not* a term but a *sentence*. As Quine points out, its translation is, perhaps, "Lo, a rabbit," or perhaps "Lo, undetached rabbit-parts." In the situation of translation, one is faced with *whole* sentences and *whole* chunks of the world, and *both* must be analyzed into their significant parts. I've argued for ways, both scientific and pre-scientific, of determining how the native cuts up her world. And these insights can be (largely) procured independently of how one translates native sentences. Although it's *not* unreasonable for the way the world is sliced up by natives to *bear on* what suitable grammatical units the sentences should be divided into, it doesn't follow, because of assumption (1), that the *semantic* units of the sentences are the same as the referential ones.

Imagine the natives are pantheists and that this belief is so ancient as to have impinged on the grammar of their language so that most of their sentences must be *translated* adverbially: "Gavagai," in particular, as "It's rabbiting" (as with our "It's raining"). *This* has *no* impact on referential facts about their terms because *we* are not pantheists and reject their ontology. We will, of course, have to talk about the things their *adverbs* are causally connected to, but this is because we take their adverbs to be the significant units of speech – at least as far as reference is concerned.[15]

I've presented a best-case scenario for learning about natives, but there are several ways that attempts to determine what natives are referring to can fail. We may be unable to get the data we need to decide among alternative hypotheses. I wrote earlier of switching items in native hoards without the natives realizing what we're doing. Such "experiments" might not always be possible.[16] Nevertheless, there can still be a "fact of the matter" about what natives are referring to in such cases.

There's another way in which *precise* characterization of what natives refer to may elude us: there really may be no fact of the matter about reference in certain cases. This is because procedures for interacting in the world are sloppy in any case,[17] and language, ours or the natives', always seems to

15　There's strain due to our translation of their language treating them as referring to only one object (the *relatum* of "It"), while we treat the causal *relata* as what we in fact see the native to causally interact with. (This is another version of the first objection posed to assumption (1).) I assume, of course, that a translation treating natives as getting things this wrong is cogent, which many disagree with. Granting this, however, assumption (1) poses a problem because referential relations construed causally come apart from referential relations construed translationally. The causal theorist either denies the cogency of this kind of translation or claims that "refers" as it occurs within the parameters of translation is not the same notion as the one pertinent to assumption (1).

16　The anthropologist or linguist tricking humans in their own habitats is, in fact, far less likely than their tricking the anthropologist or linguist instead.

17　Part II § 3.

divide the world in ways more precise than the procedures we associate with our terms can keep up with. In a certain sense, there are always referential promissory notes involved with language. Imagine we are faced with a native term we're fairly sure refers to rabbit-parts. As natives develop more refined techniques for hacking apart rabbits – we sell them metal knives that replace stone ones – we may notice that they continue to apply the same word to the new parts they are now able to chop rabbits into. This clues us into how the word can be extended – what the native term can be taken to cover.

But terminology can really outstrip mere matter here. Does the native term corresponding to "rabbit-part" cover those parts that *we* can cut the rabbit into but which the natives can't? Probably. Certainly this hypothesis is easy to test: present such a piece of rabbit to the native and see what he says once he overcomes his disgust.[18] What about those non-measurable parts we can remove from the rabbit by the axiom of choice? Hard to say. There are also puzzles about peripheral molecules hovering about rabbit-parts – how many of these should be included or excluded? Hard to say; and not just with regard to native speech.

There are limits to how far our understanding of the causal relations of natives to their world can help fix what they refer to, since we always have the problem that causal procedures do not quite divide the world neatly into what fits a term and what doesn't. The resulting slack, however, is nothing even akin to Quinean inscrutability.[19]

Quine writes that his *gavagai* example "has figured too centrally in discussions of the indeterminacy of translation,"[20] but the argument then given doesn't affect the issue of reference as posed here. Also, what he says on "gavagai" later in that article supports my interpretation of how he is undercutting the causal theorist's resources. He complains that attempts to settle the reference of "gavagai" suffer from vagueness of purpose, for

[t]he purpose cannot be to drive a wedge between stimulus meanings of observation sentences, thereby linking *Gavagai* rather to "Rabbit" than to "Rabbit stage" or "Undetached rabbit part"; for the stimulus meanings of all these sentences are incontestably identical.

18 *Contra* Quine, letting the native *see* how the part was cut from the rabbit is probably pertinent.
19 Quine (1981d) is unjustly neglected (even by Quine!). It delineates nicely the kind of problem in fixing reference by means of procedures that I am talking about here. But conceding this (and no more) leads to the position I offer in Part IV rather than anything as radical as, say, Quine's inscrutability of reference.
20 (1970a: 178). Even so, the real disagreement is that Quine thinks you can't tell how the thing is sliced by using simple ostension and querying. I agree, but claim that scientific theory and ordinary observation *can* tell us how the native slices things up if we have access to her slicing history, and some scientific theory. This makes the various options for the reference of "gavagai" purely empirical.

They comprise the stimulations that would make people think a rabbit was present. The purpose can only be to settle what *gavagai* denotes for the native as a term. But the whole notion of terms and their denotation is bound up with our own grammatical analysis of the sentences of our own language.

(1970a: 181)

Let's pick up the thread of the argument in Quine's "Ontological Relativity" (1969c). Even if we concede "gavagai" to Quine, it seems we're faced only with tame inscrutability: "gavagai" is indeterminate among rabbits, rabbit-stages, or rabbit-fusion, but not grass (because salience considerations rule that out), nor certain rocks on Pluto (for salience considerations rule that out, too). Inscrutability (with its accompanying factlessness) rampages only within fairly limited parameters.

This is *not* Quine's position. Admissible reinterpretation is far wilder than "gavagai" reveals; and unrestrained proxy functions indicate this. But before proxy functions are given full play, Quine softens us up with deferred ostension.

The idea is straightforward enough: one can indicate amounts of gasoline by pointing to a gasoline gauge, or refer to a number via a material token of a sentence of which it's the Gödel number. Anything may stand referential stead for anything else. How, then, can we possibly decide *what* is being referred to by a particular act of reference?

Quine claims our general incapacity to decide is inscrutability at second intension (1969c: 41):

> The inscrutability of reference runs deep, and it persists in a subtle form even if we accept identity and the rest of the apparatus of individuation as fixed and settled; even, indeed, if we forsake radical translation and think only of English.

But what's the argument? The alert causal theorist concedes problems with mathematical objects. But he sets such cases aside, and claims that the explanation for deferred ostension in the case of gasoline, of sub-atomic particles, or of anything else out of reach of direct ostension, is possible via the causal relations forged to such objects. (Relations, again, that are recognized not by looking at the deferred ostension situation alone, but by an examination of the whole history of our causal interactions.) And such causal connections, if required of what we refer to, do not justify the suggestion that our use of a term to refer is one that can be reinterpreted at will. It's just rhetoric to run together cases where causal resources are available with those mathematical cases where such resources are absent: it's to deny the importance of causal mechanisms in all cases of reference merely because they're absent in the mathematical ones.

If we accept the causal theorist's claim that the practice of deferred ostension, like any other referential practice, is one that we can study by examining causal relations between referrers and referred, deferred ostension poses no special problem.[21]

In any case, once Quine has the concession that deferred ostension is open to arbitrary reinterpretation (even after fixing the apparatus of quantification), a congenial setting for proxy functions is in place. For the force of causal mechanisms has been ruled out altogether. Consider, once again, "gavagai." Perhaps the term *does* refer to rocks on Pluto, via deferred ostension.

Quine's argument fits the pattern required to use proxy functions, and this despite my protests against his moves. For proxy functions come at the end of a lengthy argument (although some of the premises for that argument are supported in a different book). But Quine's "Things and Their Place in Theories" (1981c), I'm convinced, has lost sight of this, for he offers proxy functions *alone* to show ontological relativity.[22] First, he discusses abstract objects. It's noted that one could permute the entire set-theoretic universe by exchanging the referential roles played by sets and their complements, and nothing in our linguistic practice would change. One can grant the example because one has no idea how to cement words to *mathematical* objects. And although Quine concedes that there seems to be a difference between mathematical objects and concrete objects, he writes (1981c: 16): "But I am persuaded that this contrast is illusory."

Why? What follows, *by way of argument*, are permutations: of material objects with the portions of space–time they inhabit and the various permutations of a coordinate system laid on the universe. But without a justification that such permutations bear on *reference*, and no such justification is found in the discussion at hand, Quine is open to Alice's rejoinder.

---

21 Unfortunately, deferred ostension is more than the causal theorist *can* handle, as I show in § 7. Despite this, *Quine* can't use deferred ostension to establish ontological relativity because deferred ostension, as we intuitively understand it, requires largely explicit conventions: If I point to something, *A*, and by means of that, refer to something else, *B*, this only works if an understanding is already in place to enable us to so interpret the gesture. Thus deferred ostension is a poor tool for Quine, since he understands ontological relativity to apply *despite* our practices and not by means of them. No *argument* is given that his is a reasonable way to interpret deferred ostension; but, without it, one simply denies that deferred ostension can operate freely outside of explicit conventions, as Quine needs it to.

22 Quine (1986b) writes approvingly of the newer discussion as compared to that of his "Ontological Relativity" (1969c) since proxy functions are presented independently of considerations of indeterminacy of translation. But if my analysis is right, this is a mistake. What arguments Quine has to infirm the causal relations the causal theorist needs are those that occur in his discussion of indeterminacy. To present proxy functions shorn of these supporting considerations opens him to Alice's rejoinder.

§4

FIELD'S VERSION[1]

I now turn to a permutation argument in Field (1975). Field wants to establish the truth and triviality of a certain conventionalist thesis about reference. To make this thesis clear, I need a couple of his definitions. First (1975: 375), a *scheme of reference* for a language $\mathcal{L}$ is an account that specifies which words of $\mathcal{L}$ are names and what these names denote, which words of $\mathcal{L}$ are predicates and what these predicates are true of, and so on.

Next, he calls the "conventionalist thesis" the thesis that if *any* reference scheme is adequate, then *many* are adequate and, hence, that the choice of one reference scheme over another is merely a matter of convention. What's in question is an *account* of reference: the issue is not whether a different reference *scheme* could be adopted; rather, it's whether a different adequate *account* of the same set of referential practices is available, given that *one* is.

How does he show that given one such adequate account there are others? He uses a permutation argument: consider a fairly systematic permutation φ of the physical universe.[2] For *any* reference scheme S for language $\mathcal{L}$, a very different scheme S_φ can be constructed for the universe with a permutation that satisfies the same sentences. It's constructed in the obvious way: if S says a name-symbol refers to x then S_φ says the same name-symbol refers to $\varphi(x)$; if S says a predicate-symbol holds of just those things with property Q, then S_φ says the same predicate-symbol holds of those things x such that Q holds of $\varphi^{-1}(x)$.

It might seem this suggestion denies the presence of causal restraints on reference without argument. After all, utterances of a name are causally

---

1   Unlike the case of Quine, and the case of Putnam (discussed in § 5), I don't know if Field still holds the views of his 1975 publication, where I draw his argument from.

    Field (1975) apparently responds to an unpublished earlier version of Wallace (1979), where a permutation argument is also presented. But I'm concerned only with how Field (1975) uses permutations to support his conventionalist thesis.

2   "Fairly systematic," is Field's phrase. I presume the reason for the restraint is that if the permutation were unsystematic it would not yield an "adequate" account of reference, although Field doesn't say why.

connected with, by assumption, x and *not* $\varphi(x)$. Field's response to this is puzzling. He grants that terms are *not* causally connected to what they refer to (according to S_φ) but claims they *are* φ-causally connected to such objects, where φ-causality amounts to the composition of the causal relations (between us and the objects and kinds that we standardly refer to) and φ. He further asserts that an argument is needed to show the superiority of causality over φ-causality. One can grant that we need a causal theory (1975: 384) "to find the physical underpinnings of the concept of reference that people *actually employ*," but that is due to the fact that an "*arbitrary decision is already built into the ordinary usage of 'refer'.*"[3] Had we arbitrarily decided on S_φ rather than S, then a φ-causal theory would have been the suitable theory. Notice what he's claiming. Given the *account* of reference we already employ, causality is better than φ-causality. But apart from the (arbitrary) decision of having already chosen the *account* of reference we chose, this gives no reason for preferring causality over φ-causality.

Unfortunately, this is *just* a (semantically ascended) version of the simple fallacy leading to Devitt's complaint: The causal theory can't fix references of the natural-kind terms of our language because talk of "causality" is just more theory (and this theory can be reinterpreted along with all the previously reinterpreted theory). For look what happens. The causal theorist thinks that what fixes the references of terms are causal relations. *He* thinks there is a fact of the matter about whether causality or something else does this fixing. So it's no argument against him to invoke φ-causality as a possibility for our referential account. He can protest that, after all, there's no reason to think the relations that fix our terms are φ-causal ones, rather than causal ones, and it's *that*, and not the fact that causality fits in with our implicit account of reference, which makes causality superior to φ-causality.

Put the matter this way. The causal relations the causal theorist wants are ones studied in science: biology, sociology, physics, etc. In fact, the causal relations in question should *amount to* other relations we're already familiar with in other terms, relations like "touches" "sees," "hears about," etc. It's hard to see how φ-causality could amount to *these*.

I anticipate only one possible response to this objection: *of course* φ-causality *won't* amount to relations like "touches," "sees," "hears about," etc.; it must amount to relations like "φ-touches," "φ-sees," "φ-hears about," etc. We'll even have to talk about "φ-light paths."

Look where this move takes us. We're now being told to assume *without argument* that it's compatible with the causal theory of reference to reinterpret terms like "touches," "sees," and so on. The conventionalist thesis about accounts of reference has led to a conventionalist thesis about accounts of biology and sociology, too! So we're assuming *without argument* that the

---

3   Field (1975: 379).

causal theory of reference will not rely on accounts we have of resources for fixing the references of relation-terms between us and the world. But isn't *that* exactly the semantically ascended analogue of what's at issue? *Without further argument*, this last move is open to Alice's rejoinder.

One last observation about how Field endeavors to establish the triviality of his conventionalist thesis. He argues that φ-causality doesn't underwrite an account of *reference*: it underwrites an account of φ-reference. Field argues for this by means of a discussion of cases, where other interpretations for what the existence of alternative adequate accounts of reference could mean are excluded.[4] But, really, something quite simple is afoot: we're presented with a global permutation φ applying to *all* the terms of a language, including semantic ones like "reference." Such permutations show nothing without an implicit assumption that there are no requirements on a referential account to honor the *actual* causal relations underwriting the referential relations holding between terms and their referents.

---

4 That (a) *no* reference scheme is correct, or (b) semantic notions such as *reference* are instrumentally construed.

§ 5

PUTNAM'S VERSION

Putnam, it's best to immediately point out, always carefully prefaces his presentation of permutation arguments with a discussion of "operational" and "theoretical" constraints on reference. Theoretical constraints are (usually) a set of (axiomatizable) sentences fixed in truth-value and containing the terms in question. These constraints, as the discussion of the Löwenheim–Skolem theorem and the humdrum theorem in § 2 showed, can do little by themselves to fix reference.

What about operational constraints? These are assertibility conditions (however loosely) associated with our (empirical) sentences. In various places Putnam writes of the observations associated with such sentences,[1] or "experiential conditions,"[2] or (1977: 8) what can be measured or observed. The picture he gives is this: one finds associated with sentences certain test conditions, and although such test conditions are not *stipulated* to give the meaning of such sentences, how they turn out, all things being equal, constrains when we can take such sentences to hold.

At this point, Putnam's argument is very simple. Consider permutations of all our terms *that hold constant such operational and theoretical constraints.* Such permutations are easily available, and since the operational and theoretical constraints exhaust all our resources for fixing the references of the terms of the language, any position (for example, what Putnam calls "metaphysical realism") which holds that our terms refer to external objects (of whatever sort) must be wrong.

The argument applies to *all* our terms. Thus, one possible way of attempting to fix reference by the causal theory of reference seems ruled out.

---

1 Putnam writes that "if 'there is a cow in front of me at such-and-such a time' belongs to T_1 then 'there is a cow in front of me at such-and-such a time' will certainly *seem* to be true – it will be 'exactly as if' there were a cow in front of me at that time" (1978b: 126).
2 Putnam (1981: 30). This term occurs in a context where Putnam loosens the notion of operational constraint so that it will fit with Quinean strictures about holism and revisability.

For talk of causality is just more theory; we can reinterpret the term "causes" just as we reinterpret any other term. How, then, could it fix what our terms refer to, if they can't already be fixed?

Exactly here Putnam and his opponents (Devitt, Glymour, and Lewis, in particular) get into a stalemate. The first objection all of them make is that Putnam is open to what I've called Devitt's complaint, for the issue here has nothing to do with the word "cause": it has to do with *causality itself*. It is causality, the causal glue between the uses of our terms and the world, that fixes the references of our terms, and, consequently, it's that same causality that fixes what "cause" itself refers to. Putnam begs the question against the causal theorist by shifting from causality to "causality."

But Putnam is given credit for something more. He's taken to be employing a kind of sceptical strategy: the causal theorist tells a story about how our terms refer, and the causal sceptic then asks questions about how the terms for that story are referentially fixed. Glymour (1982: 177) describes it as an endless dialogue, and draws an analogy with positivistic demands for definitions. Devitt[3] alludes to the childish habit of always asking "why," and notes that the possibility of such a maneuver doesn't show the first answer was not a good one.

Lewis draws the most interesting conclusions. Apart from endorsing both of Devitt's claims about Putnam, he says that what Putnam has shown is that a "voluntaristic" theory of reference cannot be right:

> Take your favourite theory of reference. Let us grant that it is true. But let us ask: what makes it true? And the tempting answer is: *we* make it true, by our referential intentions. We can refer however we like – language is a creature of human convention – and we have seen fit to establish a language in which reference works *thus*. Somehow, implicitly or explicitly, individually or collectively, we have made this theory of reference true by stipulation. "*We* interpret our languages or nothing does" (M&R: p. 482).[4]

He continues:

> The main lesson of Putnam's Paradox, I take it, is that this purely voluntaristic view of reference leads to disaster. If it were right, any proposed constraint *would* be just more theory. Because the stipulation that establishes the constraint would be something we say or think, something we thereby add to total theory.[5]

---

3   Devitt (1983: 298–9). Also, see Devitt (1991: 227).
4   Lewis (1984: 63). M&R refers to Putnam (1977).
5   Lewis (1984: 63).

Lewis concludes:

> Referring isn't just something we do. What we say and think not
> only doesn't settle what we refer to; it doesn't even settle the prior
> question of *how* it is to be settled what we refer to. Meanings – as
> the saying goes – just ain't in the head.[6]

Lewis sees two ways out of this quandary. Either one must stick with a
causal theory of reference (*not* causal descriptivism, not the view that it is a
theory using causal terminology that fixes reference), or one must adopt elite
classes, that is, a coarser grid of available extensions for kind terms than that
afforded by set theory.[7]

I think these diagnoses of Putnam's argument are off. He's *not* employing
a sceptical strategy; nor, surprisingly, is he open to Devitt's complaint. We
can see this by simply noting how he responds to the criticisms:

> "Causal connection" is attached to R [a particular referential rela-
> tion] by causal connection, not by metaphysical glue, they write. But
> this is, in fact, just to say that R (causal connection) is *self-
> identifying*. This is to repeat the claim that a relation can at one and
> the same time be a physical relation and have the dignity (the built-
> in intentionality, in other words) of choosing its own name.[8]

What happened? No one, I think, intends to push a doctrine where something (let alone a cause) is self-identifying. Why does Putnam think they do?
The answer is that Putnam thinks he has *already exhausted* all our resources
for fixing our terms. He thinks that once operational and theoretical
constraints are honored, there's nothing else to do the job.[9]

---

6   Ibid.: 64.
7   Lewis is quite clear on the distinction between causal relations themselves being used to fix
    reference (a "genuine causal theory") and a theory with causal terms being used to fix
    reference ("causal descriptivism"). His rejection of a genuine causal theory in favor of
    causal descriptivism is not because it cannot rule out – in principle anyway – the permuta-
    tions Putnam relies on, but because he sees it as something which "often works, but not as
    invariably as philosophers nowadays tend to think" (Lewis 1984: 64). As his reference to
    Unger (1983) shortly after this sentence makes clear, what he means by "works, but not ...
    invariably" is that a genuine causal theory of reference does not fit well with intuitions
    many have about certain thought experiments.
8   Putnam (1984: 7). Also see Putnam (1983a).
9   I've found only one piece of text where Putnam explicitly says this:

> It is not that there aren't various naturalistic connections between the word "refer-
> ence" and R_{17} [one of the many referential relations possible via permutation]; it is the
> idea that one of these *declares itself* to have the honor of making R_{17} *be* the relation

Devitt and Glymour engage Putnam at the wrong point. They offer the causal theory to Putnam *after* he has honored all the resources he thinks causal theorists are entitled to. So Putnam is not open to Devitt's complaint. Instead his error is similar to that in Quine's *Word and Object* (1960): Putnam has surreptitiously infirmed the resources the causal theorist hoped to have – and he also does it early on, when discussing so-called *operational constraints*.

A causal theorist must take operational constraints seriously, because they contain the causal relations needed to affix terms to the world. But, in general, such constraints are *not* observational, or at least, not purely so, as Part II § 5 showed. A practicing physicist may interact causally on a daily basis with sub-atomic particles using devices designed especially for such purposes. At the proximate end of the causal chain, "observational" predicates apply, but at the other end they don't. Operational constraints, on the causal picture, cut deep into the world – they don't stop only at what we "observe," and, consequently, to honor their restraints we must accept that they fix more than appearances.

Putnam's operational constraints *do* (largely) stop at the observational: indeed, at times he uses the phrases "operational constraint" and "observational constraint" *interchangeably* (as in Putnam 1989: 215). The irony is that Putnam usually takes care to note in good Quinean fashion that a strict distinction between theory and observation isn't available. But he fails (like his critics) to realize that it's here, in restricting the operational resources available to the causal theorist to whatever is (more or less) observational, that he begs the question against that theorist.

So, if I'm right, Putnam and (most of) his critics talk past each other. Each time Putnam claims that "cause" is just a word, everyone thinks he's opening himself to Devitt's complaint, when Putnam is noting only that nothing more is left for that word to pick out that hasn't already been accounted for. The diagnosis: the disastrous *exposition* of the argument against the causal realist that puts permutations center stage.

Perhaps it can be argued that the appeal Putnam finds in permutation arguments is due to an unconscious operation of the error leading to Devitt's complaint.[10] In any case, although I can find clear examples of the error leading to Devitt's complaint in the later Quine, and in Field (1975), but *not* in Putnam, Putnam *alone* is widely believed to be subject to it.

---

of reference *independently from all operational and theoretical constraints* that is entirely unintelligible.

(Putnam 1983a: 296)

Only the last set of italics are mine.

10 Actually, Putnam's concern with the "normativity" of reference suggests something else: he senses detachment of the referential idioms from the causal relations that underwrite them, and it's this that motivates the appeal of permutation arguments for him. More on this later.

Recall what a legitimate attack on the causal theory of reference using permutations looks like: it should directly oppose the resources the causal theorist takes himself to have, and show they can't do the job he wants them for. None of the arguments canvassed thus far do *that* job. Instead, they cut out those resources in a way that only an already intimidated causal theorist would accept. Interestingly, Putnam *has* offered a couple of arguments against the causal theorist that really do engage him in a non-question-begging way, although, peculiarly, these arguments are offered by Putnam as subsidiary to his model-theoretic arguments.[11] I take them up in the course of § 6.

Two last comments. First, I've argued against Putnam's version of the permutation argument by claiming that his operational constraints are too weak – that they are, pretty explicitly, observational constraints only. What if we strengthen them? What if we include everything that the causal resources available to speakers buy them? Unless these resources can underwrite our referential practices entirely – and of course I've already hinted fairly strongly several times that they won't – won't certain permutations still be possible? Won't, in other words, a version of Putnam's argument be *right*?

Maybe. But it's the differences that count. Putnam says nothing about exactly how far the causal resources available to us to underwrite reference fall short – and he doesn't because of his implicit identification of operational with observational constraints. This results in a picture where the permutations apparently available are so global in nature, so transforming in their semantic implications, that there's no point in attempting, say, to see where our metaphysical commitments involve a causal connection between things out there and us, and where such commitments go beyond such causal connections. Indeed, Putnam has stressed many times that it's idle to attempt to separate out those aspects of our conceptual scheme that are "made up" from those aspects that are not.[12] A view that starts from the acceptance of operational constraints on reference which are causal in nature and, consequently, cut deep into the world (metaphysically speaking), yields a view rather different from Putnam's, as I show in Part IV.

Second point. I hope the reader noticed the analogy between the lesson that "Putnam's paradox" is, on Lewis' view, supposed to teach us and the

---

11 This is not quite the case in Putnam (1992). There the issue of the interest-relativity of causality takes center stage, and permutation-style arguments are relegated to a footnote, although they are still described with approval. It is also not quite the case in Putnam (1978a). But there the interest-relativity is used rather directly as a gloss on Quine's arguments for indeterminacy. The version of the permutation argument that Putnam adopts in the final article (1978b) in *that* book (1978a) doesn't deal with the interest-relativity of causes.

12 "The Trail of the Human Serpent is Over All" is the title of a sub-section in Putnam (1987) which presents an argument for conceptual relativity – the claim that there are many equally acceptable ways to slice up a world (see pp. 16–21).

lesson we were supposed to have learnt about conventions from Quine's "Truth by Convention" (1935). In both cases, a certain sort of convention-alism that involves acts of stipulation is supposedly refuted because it's seen as open to a kind of regress argument: for Quine, in order to stipulate that certain logical idioms have certain meanings, one must presuppose those very meanings. For Lewis, in order to fix the references of our terms, we have to "say or think" the constraint needed, and in so doing presuppose the very sort of fixation of our terms that we are trying to establish this way.

The objection fails (in both cases). In brief,[13] it assumes that a stipulative act requires a meta-act of the same sort which establishes that the act in question fits the rules established by the stipulative act itself. But this is false. We can recognize, posthumously as it were, that our stipulative act has the right properties; in this particular case, we can obey a constraint on refer-ence without actually having formulated at any time what that constraint actually is.

I did say Lewis' gloss of Putnam's argument is the most interesting. Here's an indication why, although the full picture must wait until Part IV. If one accepts that it can't be that every referential stipulation we use is explicit, then something other than such conventions must be fixing refer-ence in such cases – or nothing is. As Lewis sees things, either elite classes do it, or, with Putnam, nothing does.

But there's another possibility: tacit "natural" causal relations do the job. For example, except under unusual circumstances, pointing at something *is* referring to it.

This suggestion seems ruled out because of "normative" intuitions we have about reference. Reference does not have to arise from such simple rela-tions. This is one way to understand Quine's point regarding deferred ostension, and it is also how we can understand passages such as § 74 of Wittgenstein's *Philosophical Investigations* (1953).

However, a point needs to be repeated. That reference *need not* be consti-tuted from such simple causal relations does not mean that it *is not* so constituted. And whether it is or not may be something we can determine. Furthermore, the intuitive examples used to show cases where reference *is not* constituted by natural causal relations involve *explicit* conventions. Further argument is needed to show that "non-standard" referential rela-tions could exist without such explicit conventions being in place.

Here's an illustration. Imagine we have a group of individuals who already react fairly similarly to stimulations. In this case pointing can arise as a practice (*without* explicit conventions) simply because when someone makes a gesture of a certain sort, everyone (more or less) notices the same thing. Once something like this is already in place (as bedrock), conventional

---

13   In brief, because I discuss the matter at length in Part 2 § 5 of Azzouni (1994).

modifications of the referential practice can be introduced. We can *now* decide (as a group) that pointing means *this* rather than *that*.

If the members of a group do not have a tendency to respond similarly in certain respects, the practice won't get started to begin with. And if a different group (of humans, aliens, whatever) naturally respond *differently* to a gesture in the presence of something, then the causal relations on which reference will be built *must* differ, even if the referential relations themselves are the same.

The story I have just illustrated must illuminate the strong impression that we *can* make any symbol mean whatever we want. Part of what's needed to do this is already in place: we can make anything *mean* whatever we want only if (for the most part) we do it explicitly, starting from a basis of what we tend to do naturally (for without such a basis there is no bedrock for a language to get started on in the first place). And this points to what we do when we meet an alien group and wish to interpret them. Even if they defer ostension all the time, even if their explicit referential conventions are rich and complicated, we are not (in principle) lost. For a genetic story is (in principle) available. Start with what they do naturally, if we can figure it out, and go on from there.

§6

THE ONTOLOGICAL STATUS
OF CAUSALITY

In the course of Part III, I've incurred a couple of implicit promissory notes, some of which I'd like to discharge now. Although ultimately I'll argue that the causal theorist cannot have what he craves – causality is not a relation that underwrites reference – still, as the concluding paragraphs of the previous section should have made clear, the causal idiom is a robust and important one which plays a significant role in reference, at least when the objects referred to are certain empirical ones. To make this claim plausible, it's not enough merely to have fended off permutational threats. One must also say a little about how the causality relevant to reference operates, and how we gain access to it. This is, in large part, the topic of this section.

I start by observing that many of the causal relations pertinent to reference are, pretty obviously, observational regularities described in ordinary language. Recall how Kripke elicits intuitions to support a causal interpretation (for how the reference of a name in certain cases is passed on) as opposed to a descriptive interpretation:

> Someone, let's say, a baby, is born; his parents call him by a certain name. They talk about him to their friends. Other people meet him. Through various sorts of talk the name is spread from link to link as if by a chain. A speaker who is on the far end of this chain, who has heard about, say Richard Feynman, in the market place or elsewhere, may be referring to Richard Feynman even though he can't remember from whom he first heard of Feynman or from whom he ever heard of Feynman. He knows that Feynman is a famous physicist. A certain passage of communication reaching ultimately to the man himself does reach the speaker. He then is referring to Feynman even though he can't identify him uniquely.
>
> (1980: 91)

And later he writes:

> In general our reference depends not just on what we think ourselves, but on other people in the community, the history of how the name reached one, and things like that. It is by following such a history that one gets to the reference.
>
> (Ibid.: 95)

All the literature on this topic, despite disagreements about how to sort out intuitions about reference, and regardless of whether a purely causal approach or one combined with descriptions in some way is taken, discusses causal resources that can be pointed to without the use of sophisticated instruments or scientific theory.[1]

At this point, then, the pertinent discussions of the epistemic and onto-logical status of observational regularities raised in Parts I and II can be brought to bear on the epistemic and ontological status of these causal rela-tions; and in this way the claims made there about observational regularities can be used to justify (to some extent) theories of reference that depend on causality in a serious way. For I can, this way, defend them against objec-tions which turn on denying the legitimate status of such causal resources.[2] As an illustration (more will follow), consider the claim that the causal rela-tions important to reference are entirely theoretical ones. One does not observe causes at work; so, the suggestion goes, one theoretically imposes them upon what one observes.

Let's not understand this objection in a Humean fashion, that is, as the claim that all that can be observed by us are constant conjunctions, and that the necessity, or law-likeness, of such conjunctions is imposed. That may be true for all I care here,[3] for the suggestion that has to be rebutted is not this, but the more specific idea that the notion of "cause" or "regularity" is a purely *theoretical* one, and consequently that causes are events which are not

---

1　See, for example, Evans (1973) or Unger (1983). I should add that it's certainly not true, when it comes to kind terms anyway, that the causal resources utilized are restricted to those arrived at without the use of sophisticated instruments or scientific theory. Indeed, one can cook up examples so that this need not be true even with certain names (although in practice it usually is). The point I'm trying to make in the text above is that causal rela-tions are a rich class of events that we gain access to in many ways, and that one of these ways is *observation*, as I understand it.

2　Since, in some sense, the causal theory of reference is going to fail anyway, why *bother* defending it? Because it's extremely significant *how* it fails: how it fails points to what the right view of reference should be. In particular, the burden of this chapter is to show it doesn't fail *because* the *causal relations* the causal theorist wants are ontologically suspect in a way that makes them useless for his purposes.

3　Actually, I think something like this *is* true. For a gross regularity may be a regularity that can be hedged into, or derived from, a law, or possibly not. Which way it can go turns, as I've argued in Part II § 4, on how the gross regularity gets incorporated, if it gets incorpo-rated at all, in scientific doctrine.

observed in any sense. As an illustration, there is the Davidsonian claim mentioned earlier (III § 3, note 7) that reference is a theoretical relation that arises only within the context of the imposition of a truth-theory on a language, and consequently cannot be used by the causal theorist to fix reference as he would like. The causal theorist can reject this claim by again pointing out that the causal relations in question (e.g. ones pertinent to the reference of most proper names) are observational regularities.[4]

Consider the similarly motivated worry that one is unable to supply necessary and sufficient conditions for acts of reference in physicalistically acceptable terms; and so the referential relations the causal theorist wants are not kosher, scientifically speaking. This objection, too, can be defused by pointing out that the causal relations in question are gross regularities that naturally admit of exceptions.

The causal theorist also avoids the requirement that reference be definable in physicalistically acceptable scientific terms by noting that gross regularities are, for the most part, epistemically independent of the empirical sciences. As with most core terms, "refers" may pick out the appropriate causal relations without there being a definition of "refers" available in more pristine (causal) terminology that can do the job.

This last rejoinder won't satisfy anyone who is worried, not that necessary and sufficient conditions for "refers" are unavailable in scientific terms, but that such necessary and sufficient conditions are unavailable even with regard to causal talk which involves core terms and gross regularities.[5] This problem, which I've raised earlier (when I described the first objection to assumption 1), I'll call "the problem of the normativity of reference," and I'll again put off explicit discussion of it until later.[6]

Now consider two objections raised by Putnam against the causal theory of reference – the one in passing, the other repeatedly – both of which draw their inspiration from an attempt to deny to causality sufficient robustness for the causal theorist to take it as fixing the reference of empirical terms.

The first argument is in Putnam's essay "Beyond Historicism" (1983a: 295–6). It raises the concern that causality must link "cause" itself to its referent:

> This is flagrant violation of the theory of types. There is, in fact, no relation which contains in its extension an ordered pair one of

---

4   He can do better: He can support assumption 1 *and* assumption 3 by observing that it's not just observational regularities about *causes* that philosophers point to in this literature, and that ordinary people clearly rely on; these are simultaneously observational regularities about reference *too*.

5   I see Blackburn (1988) as worried about this issue, for the causal relations he concerns himself with are entirely observational ones; I see Putnam (1992) as partially worried about this too.

6   See the next section, and Part IV.

whose members is that relation itself (i.e., which is such that something bears that relation to the relation itself).

Weirdly, Putnam claims we could change our logic to get around this, but logic isn't pertinent; and the matter is more delicate than he indicates. The problem is this: we *could* change our set theory and go non-foundationalist,[7] or perhaps we could shift from talk of reference to talk of a hierarchy of referential relations. But, recall, the causal theorist wants "reference" to instantiate a natural kind (perhaps even a physicalist one). Field (1989: 211) has noted that the most widespread conception of physical properties and relations is a predicative one. I don't want to say outright that this means the causal theorist is in trouble; but how he's supposed to approach this problem is *not* obvious to me.[8]

The second argument has been emphasized more by Putnam, at least in sofar as it features more often in his writings than does the first argument.[9] It's quite simple: the ordinary use of "cause," and the one that the causal theorist is going to need, presupposes a distinction between causes and standing conditions. But such a distinction, Putnam claims, is related to *explanation*, and consequently is interest- and salience-relative. Furthermore, this shows that the notion is not identical to the notion of a contributory cause (not every contributory factor to an event is a "cause" in the sense that we use the term in ordinary language). Putnam illustrates these points, which he attributes to Hart and Honoré (1959), by means of a couple of examples.[10] Here's one: if a pressure-cooker has a stuck valve and explodes, we may say the stuck valve was a cause of the explosion, but we would never say the absence of holes in the body of the vessel caused the explosion, despite the fact that the absence of other holes in the body of the vessel is *a* reason why there was an explosion, and, moreover, despite the fact that a stuck valve just amounts to the absence of such a hole.

---

7   For a nice introduction to non-foundational set theory, see Aczel (1988). For applications of it to non-predicative semantic phenomena, see Barwise and Etchemendy (1987).

8   I guess I really do think these considerations show that the causal theorist *is* in trouble, but only because it makes clear in a particularly neat way that the constraints on reference and the constraints on causality do not fit very well together. For consider. The problems with reference facing causal views here are quite analogous to the problems that arise regarding truth (and that should be no surprise, since truth and reference keep such close company). Self-applicable concepts breed puzzles in any case, but reference as an empirical natural-kind seems to face even worse troubles, since it's hard to see how it's supposed to be self-applicable in the first place. A causal hierarchy, for example, generated in analogy to the Tarskian hierarchy does not fit well with our views of how causality operates in the world. Instead of reducing reference, in some sense, to causality, this suggestion would really elevate causality to a full-blooded semantic notion.

9   Putnam (1983e, 1984, 1989, 1992).

10 Putnam (1992: 47–8). See also his (1978a).

The notion of cause needed here, therefore, is one largely couched in ordinary language, and is context-sensitive and interest-relative. Putnam clearly thinks the causal relation therefore must be a product of our theorizing in the sense that it cannot be a robust relation between us and items independent of our theorizing.[11]

The argument won't do. As many have observed, just because our *choice* in describing *A* as a cause is context-dependent and interest-relative, doesn't mean that *A*'s *causal role* (among its fellow-causal-agents in the nexus of causes) is context-dependent and interest-relative. *A*'s role can be as absolute as you please. Thus our reasons for singling out aspects of the total cause, although epistemically and pragmatically significant, have no bearing *per se* on the ontological status of what we have singled out. The fact that our interests and the context make us single something out doesn't mean that it isn't out there *already*; it might not be, of course, but this is hardly shown by pointing out that with other interests, and in another context, one might not notice it.

Now I've already conceded that causes of this sort, like anything else observational, are ontologically suspect, in the sense that how, exactly, they are to be embedded in scientific doctrine is up for grabs.[12] But this is hardly because of their interest-relativity and context-sensitivity. In fact, it's in the contexts where such regularities operate that they're most likely to be respected by science *as they are*; for it's in virtue of how they're to be extended, what their scope is, that they're most open to modification when science progresses, as opposed to contexts where such regularities are well-established.

A necessary condition for some phenomenon being taken as robust, scientifically speaking, is *not* its being entirely subsumable under some law; nor does it consist in the ready provision of definitions for it. Rather, all that's sufficient is that it become the subject of intense scrutiny on the part of scientists. At the end of such a (possibly long) process, the subject matter *may* allow itself to be subject to law and definition, but its failure to do so casts no ontological aspersions upon it.

---

11   Putnam (1992: 57–8) writes:

> I have argued that the notion of causation … has a cognitive dimension, even when we use it in a statement about inanimate objects, for example the statement that the stuck valve caused the pressure cooker to explode. The cognitive or "intentional" dimension lies in part in the presupposition that hearers of the statement regard such facts as that the vessel of the pressure cooker does not have a hole in it as "background conditions" which may be taken for granted, as well as in our knowledge of the salience that others attach to the condition of the valve.

12   Indeed, this is so in the strong sense that, ultimately, they might be explained in ways that make the term "cause" inappropriate, e.g. quantum mechanically. This is subject, of course, to the issues of tractability raised in Part I.

Causality is *no* exception to this. Philosophers may look askance at causality because of difficulties *they've* had supplying definitions and suchlike for it, but its robustness from the scientific point of view is amply testified to by the intense scrutiny to which it is subjected. What I called the technology of epistemology in Part I has to do with the practice in empirical science of carefully refining the causal connections between us and the objects we study; this is a practice in ordinary life, too. I illustrate this point with a couple of examples.

In Strasberg's entry "Acoustical Measurements" in the *Encyclopedia of Physics*, we find an explicit discussion of a number of measurement tools (various kinds of microphone and hydrophone, the Rayleigh disc, fiber optics, and so on), their range, sensitivity, and other pertinent properties, in the specific context of physical theory. That is, a certain group of causal relations between us and certain phenomena (in this case, fluctuations of pressure, temperature, density, etc. of matter) are studied.

A mark of the study of the means of access to something is a concomitant concern with the artefacts of that means of access, as I noted in Part I § 8. This sort of concern is commonplace in the sciences, and in ordinary life, too. In the latter case, we're so habituated to adjusting our information gathering to the avoidance of artefacts that we often don't even notice we're doing so. Consider "floaters," small translucent or opaque items that sometimes float across one's vision. Even without any idea of what these things are (*retinal detritus*, I was mortified to learn), one is pretty good at recognizing that they're not something out there (e.g. Casper the friendly ghost) but rather are a product of events in the eyeball. No different in principle are the considerations that lead physicists to regard most of what's seen through the microwave-sensitive eyes of the COBE satellite as instrument artefacts.[13]

A scientific study of a causal means of access is one that can take place fruitfully even if there is no adequate theory of how that means of access operates. Consider vision again. Despite the absence of a (complete) theory of how human vision works, Weiskrantz and his associates still had enough of a grasp on the causal means by which sighted adults use vision to screen out possible artefacts[14] in their study of the "blindsight" of their patient

---

13 This is a really interesting example. Scientists used a statistical analysis to show that the global properties of the map designed on the basis of the COBE satellite are right even if it's not possible to point to a particular item and say: "That, at least, is not an instrument artefact." See Powell (1992) for a readable discussion of these matters.

14 "Artefact" is used interchangeably in two ways in the above examples. Sometimes the artefacts in question are "perceived items" which are products of the perceptual apparatus rather than of the world at large. At other times they are results in the experimental situation apparently due to one phenomenon, but actually due to others. "Artefact" in both senses is pertinent to the point being made.

DB. In particular, they were concerned to rule out "possible artefacts due to stray light or other cues,"[15] "adventitious eye movements,"[16] the "diffusion of light within the eye," as well as "artefacts due to reflections among surfaces in the room."[17] What's called for is not a *complete* theory of how the causal connections made here operate, anymore than this is called for in any other branch of science. Rather, all that's needed is enough of an understanding of the apparatus used (in this case, light, eyeballs, the brain, the visual environment) to enable the researchers to exclude certain possibilities.

Two last observations. First, for contrast, notice the complete absence of this sort of practice in the mathematical sciences. It is this that makes classical Platonism so implausible, and why perceptual analogies, in particular, are such poor ones to use when explaining how we grasp mathematical objects. For however we grasp mathematical objects, it cannot be by anything akin to perception or intuition, as there's nothing corresponding to the study of the process of the recognition of mathematical objects, or an account of how to recognize the artefactual by-products of any such process on the part of mathematicians.

Second, in describing the focus on causal relations (both in scientific and ordinary contexts), when they're used as a means of access to something, the reader perhaps could not help noticing that the general framework in which these considerations arise is epistemic rather than referential. The concern on the part of scientists, and ordinary folk, is with what we know or fail to know about the objects in question, and not about how well we have succeeded in referring to them. This might suggest that proponents of the causal theory of reference are making some sort of conceptual error: causal relations, intuitively speaking, bear on *knowledge*, not on *reference*. I explore this worry further in Part IV.

---

15   Weiskrantz (1986: 35).
16   Ibid.: 41.
17   Weiskrantz (ibid.: 42). The avoidance of artefacts in this sense is a theme of this monograph. The issue comes up repeatedly throughout Parts 2 and 3, and a number of experiments are designed with the specific aim of ruling out one or another artefact. See e.g. Chapter 10 (pp. 90–2).

§7

SOME PUZZLES ABOUT
REFERENCE

Until now I've tried to accommodate the causal theorist: I've tried to show why certain arguments against causality's major role in fixing reference are misguided. Nevertheless, there's a (*prima facie*) serious problem with the idea that causality does underwrite reference, although it's not quite the problem raised by permutation-plyers. The problem is only *prima facie* because it is drawn from intuitions we have about reference, and intuitions can sometimes be explained away. In any case, I bring attention to a set of thought experiments that seem to show it's not necessary that a causal mechanism directly underwrite the referential relation.

Interestingly, this set of thought experiments is close enough to what permutation theorists have argued for that it should be no surprise it has not come up in the literature earlier; although simultaneously, it's different enough that the philosophical implications that can be drawn from it are rather distant from implications drawn by permutation theorists about the irreality of reference.

Consider a permutation φ of the physical universe which shifts every space–time point some large number $\aleph$ of light years in some fixed direction Δ. I don't care exactly what $\aleph$ and Δ are, provided they're chosen so that every space–time point within our light cone is translated out of it. This is to guarantee, according to contemporary physical theory, that there are no causal relations possible between us and the translations under φ of space–time points we're in causal contact with.

Some may fear that what we're in causal contact with, in any case, are not space–time *points* but chairs, tables, various animals, and the like, and that there is no obvious connection between *these* things and space–time points. No problem: understand φ so that it maps such objects to their translations $\aleph$ light years away in direction Δ. Talk of space–time points may be seen as a crisp shorthand for describing φ in a way that makes it clear what every object (that we *can* refer to) is mapped to under it.

Now I'll make an important claim, and spend some time defending it. It seems perfectly clear, intuitively speaking, that we, as a collective of referers, could *explicitly* decide to refer, from now on, not to the objects we normally refer to, but to their translations under φ. This possibility should be shocking to causal theorists of all kinds, for if we adopt this proposal, causality would no longer underwrite the referential relation to (pretty much) *anything* we normally refer to.[1] I call a language whose references have been shifted in this way, by a permutation φ, a *φ-language*. Here are several remarks about φ-languages.

First, the reader should not be tempted to revive Field's suggestion that the appropriate *account* here is that what's fixing reference in φ-languages is φ-causality rather than causality. For although φ-causes are now the referents of our term "cause," this shouldn't fool anyone into thinking that φ-causality is fixing the referents of our terms; for *that* move is open to Devitt's complaint. The confusion in this case would be one between the theory that we have (which is identical, syntactically speaking, to the theory we had before) and what is picked out by that theory. The causal theorists that I'm speaking of are those who think that it's not a theory of causality but *causality* itself that fixes (or helps fix) reference, and such causality is patently absent here.

Second, the reader shouldn't think this kind of case is the sort of thing permutation theorists have been arguing about. As I've mentioned, the intuition being elicited here is that the shift in reference contemplated is one that could be adopted *explicitly* – it's not being suggested that in fact objects under the permutation φ are ones we could actually be referring to before making the change to a φ-language (i.e. reference is inscrutable); nor, as I stressed in the first observation, is it being suggested that the change in language contemplated is an idle conventional shift in our *description* of how we refer.

Third, the causal theorist, of whatever stripe, is sooner or later going to have to grapple with the scarcity of the causality needed to glue our terms to the universe. This glue *is* available, in the sociological realm as it were, for proper names such as "Richard Feynman." And glue is available for certain samples of kind terms (although, generally, not for *every* instance of a kind term).[2] We've interacted with (modulo the linguistic division of labor, anyway) *instances* of giraffes, planets, stars, gold, and so on. And we also have, at least at the proximate end of our range of measurements, successfully operationalized the measurement of certain quantities. But the causal picture, however understood, has to include the fact that reference extends far beyond such causally tame cases. φ-languages owe their existence to our capacity to use this non-causal kernel contained in any language, rich enough to refer to the sorts of things our language can refer to, to define a permutation which results in *none*, nearly enough, of the ordinary proper

1 On the parenthetical qualification "pretty much," see note 3.
2 This point will be elaborated further in Part IV.

names and kind terms in the resulting φ-language having any causal connections to what they refer to.[3]

The *fourth* observation is one I'll spend a little time on: the causal theorist might want to argue that the φ-language contemplated is not a language which could be learnt by anyone, or even by a community; for the learning of any language, so the argument would go, must start with ostension, and this is not possible here. I cannot respond to this objection by saying that the language with its standard references is identical, as far as users are concerned, with the new language, because they're *not* identical on the view I've been arguing for. And in fact the causal theorist is entirely right – in this respect at least: one could not quite learn the new language from scratch, as it were, the way one learns the language we currently use.

This concession won't help the causal theorist very much, for remember what's being argued for. The claim is not that causality isn't needed in some way to learn a language; that's quite true. What's being argued for is that it's intuitively possible to have a language where the pertinent names and kind terms are *not* linked causally to their referents. This point wouldn't be disturbed by the fact, if it were a fact, that such a language couldn't be learned from scratch, but only parasitically, as it were, on another language the referential relations of which *were* composed, in some way, of causal relations. One can certainly imagine situations, that is, where one first learns a language $\mathcal{L}$ the ordinary way (via ostension), and then learns a second language $\mathcal{L}^*$ (which one henceforth uses exclusively) by referentially piggybacking via a permutation on $\mathcal{L}$.

As it turns out, a φ-language is *not* a language we must learn on the heels of an antecedent pre-language. And this is because of the linguistic division of labor. Imagine a society, call it Schmindia, where there's enormous respect on the part of the general population for a class of linguistic Brahmans. Pretty much whatever the Brahmans claim is assumed true by the rest of the population.[4] The Brahmans have discovered set theory, and understand it in an ontologically classical way (sets are *not* in space–time, etc.). Furthermore, there is a ritual of passage which all young Brahmans go through when they reach adulthood. Aside from the usual things, tatooing and all that, they are told what the terms of their language, the language that *everyone* speaks, *actually* refer to: terms refer not to the objects or groups of objects they've always *seemed* to refer to; rather, the proper names

---

3   "Nearly enough" because some of our measurable quantities, like inches, are still linked to causally available items. Also notice, in any case, that I have *not* claimed that none of the *terms* (apart from ordinary names and natural-kind terms) of the resulting φ-language are causally linked to what they refer to, because that's not true. See the *fifth* observation, below.

4   There can be disagreements among Brahmans, of course, but these are ironed out behind closed doors in a collective and semi-judicial way.

refer to the unit sets of such objects, and the kind terms to groups of such unit sets.[5] So, the term "green thing" holds not of green things but rather of unit sets of green things; so, too, "Schbuddha" refers not to the actual man (however extended in space–time – or out of it – he might be), but to his unit set.[6]

The novice is taught that she has been speaking this language *all along*. It is not that she is initiated (in part) by being taught a secret code, a homophonic copy of the home language with esoteric references; no, this *is* the language of the vulgar. The Brahmans say regularly of non-Brahmans that "they know not of what they speak," and due to the division of linguistic labor (that the population will yield to the decisions of the Brahmans in respect of what their terms refer to, just as we do with professional scientists), Brahmans speak the truth.[7]

One might well worry about how the intuitions raised by this example, and others easily designed along the same lines, fit with the intuitions raised by Kripke and Putnam to argue against description theories for reference. Consider, for example, the following. Intuitions about proper names seem to suggest that a proper name *is* causally linked to what it refers to (at least in certain central cases). *Our* intuitions, I think, do not allow any group, even scientists, to re-route reference in these cases except by considerations that show the causal chain underwriting reference is not what people thought it was. An archaeologist, for example, might be able to unearth data showing that "Aristotle" does not refer to the man we thought it referred to; but nothing any scientist might unearth could upset our linking terms to objects by causal chains of this sort.[8]

I think matters sort out this way: although we can see what the linguistic habits (and intuitions) of members of a society such as Schmindia would be like, we can also recognize that intuitions we have about reference

---

5   This is *not* a φ-language. But the example is easily modified so that it is a φ-language the young Brahman learns she has been speaking, and the lessons the example teaches will stand. I'm just not in the mood right now to attribute ambitious astronomical beliefs to Schmindians.

6   We can imagine this referential shift to be accompanied by a peculiar (from our point of view, anyway) hygienic doctrine: unit sets of objects are "clean" in a way ordinary objects themselves are not. Naturally such false views about the properties of what is referred to do not affect the force of the example.

7   Those inclined to disagree with this conclusion face the problem that *anyone* in Schmindia, if presented with this thought experiment, would agree with my analysis. On what grounds, then, are we to say that *they're* wrong?

8   For cases like "Aristotle" this is fairly clear. There are problematical cases, of course, where terms have clearly shifted in reference – a causal chain, that is, has been repudiated. See Evans (1973: 11) for his "Madagascar" case. But such shifting shows even more of an alliance, as far as our intuitions are concerned, with reference as it is practiced by the "vulgar."

presuppose that our own society is not like that. We would repudiate attempts on the part of any group which tried to control reference the way the Brahmans control reference in Schmindia. Intuitions about causality, therefore, are not intuitions which bind *all* possible referential practices; they are not, that is, *necessary* constraints on reference, nor are they seen intuitively as such, as the intuitive plausibility of the referential practices in Schmindia makes clear. Rather they are indicators of how we take referential practices in *our own linguistic community* to operate.[9]

Fifth, the causal theorist might argue against φ-languages this way: one advantage, or so one would have thought, of the causal theory of reference, or of theories in which causality plays a major role, is that the causality involved helps explain how we refer to what we refer to, and furthermore, of how we learned to refer in the first place. Such an explanation seems absent in the case of φ-languages; for providing an explanation of this sort that *is true* turns crucially on talking about causality, and we cannot refer to causality in a φ-language with the term "cause."

The final clause of the last sentence is right; but that doesn't necessarily pose a problem about providing the requisite explanation, because if the original language is rich enough to define φ the resulting φ-language is usually rich enough to define φ^{-1}. If we have φ^{-1}, in turn, it's easy to define a term that refers to causality (not φ-causality), and use *that* to explain both how we refer to the things we refer to via the φ-language we now speak, and how we learned the φ-language itself in the first place.[10] (A slogan: *You can kick away the ladder, but don't if you need to show someone how you got up there.*)

Sixth, φ-languages fail to show what the permutation theorist wanted to show. These examples all turn on "voluntaristic" intuitions about reference (albeit those, as Schmindia showed, which respect Putnam's linguistic division of labor); also they rely on *explicit* conventions. So they don't offer direct lines of attack for referential irrealists who wish to deny the role of causality in fixing reference. Indeed, the causal theorist can respond to my examples by making, perhaps, only a minor retreat. He can point out, as I've noted, that although causality is not a *necessary* constraint on reference, nevertheless, it's a constraint *among us*. And perhaps that is all that's needed for a significant analysis in naturalized semantics. Never mind all the possibilities imaginable: reference, in *our* niche, is built on causality, and the intuitions about reference that Kripke and Putnam raised in their classic discussions show this.

---

9   Do these presuppositions at least make causal constraints on reference (contingently) *a priori*? No, for it is not *a priori* that we do not actually belong to a Brahman society.

10   The Schmindian dialect does *not* confer this ability on its speakers, although given the hygienic prejudices of the Brahmans, this is perhaps no drawback for *them*.

His second point is this: even in (pathological) cases, where causal relations do not underwrite the references of proper names and kind terms in a direct way, causality is still needed to explain how we refer in the first place, to explain, that is, how speakers were able to learn the languages they speak. Causality has still a central role to play in reference; it's just that this role has turned out to be more subtle than anyone expected.

I'll show in Part IV that the sanguine attitude just offered on behalf of the causal theorist is not justified. Even with regard to our own language, causality does not quite play the role the causal theorist hopes for. Moreover, the existence of φ-languages shows that we must ultimately develop a picture of reference akin to the picture of truth sketched in Part II § 7.

Finally, I've assumed all along that φ-languages could be explicitly adopted. But *is* explicitness a requirement? It would seem so: languages with referential schemes based on what we do *naturally* don't require explicit conventions, since we'll do what they require "naturally" (indeed, this applies to any practice based on what we do naturally). Notice an important fact: what is done naturally is something that must be done by everyone (pretty much) naturally, otherwise the needed social coordination of the practice is lost. This seems to suggest that only languages with terms fixed by causal relations used in a rather direct way (as traditional causal theorists hoped) are ones that we could adopt without explicit decisions, since only these resulting referential schemes are natural.

Here's a possible counter-example to this thought. Imagine that a community, which holds its traditions in enormous respect, makes the discovery (perhaps in the form of a collection of ancient tablets) that their language, in the hands of their forefathers, was Schmindian in referential import.[11] Do the terms, as currently used, refer to unit sets?

They do if the community *takes* them to. Although I can easily imagine communities that would react pretty much the way we would if we'd made such a discovery: "How quaint!" or perhaps (less respectfully) "How weird!" I can also imagine communities where the "fact" about referential practices on the part of the forefathers would be accepted without qualms, or where such a discovery would lead to disagreements (a civil war?): it all depends on how slavish the community is towards forefatherly practice. In each case, however, or so it seems to me, how the community goes dictates what their terms refer to.

If this is right, then a community could have an implicit set of referential practices where (a) their language was a φ-language, and (b) this fact was not known to them. Although the cases in question allow the use of a φ-language on the part of a community to be implicit, this use must still trace

---

11 Again suppose that referential practices of the forefathers were accompanied by, say, hygienic doctrines which are still taken seriously.

back to an explicit forefatherly decision (or something similar). Thus the counter-example is a counter-example to the claim that only languages the references of whose terms are fixed by "natural" causal relations can be languages used without explicit conventions. This claim is qualified only in so far as diachronic applications of the linguistic division of labor can be applied, and in so far as "we language users" don't include the forefathers who explicitly adopted certain conventions.

Although I have not done much to circumscribe the options here (I'm not sure I can), it's clear that questions raised about what our terms refer to, in a broad way anyway, are questions that we take to be determined by the conventions we have either implicitly or explicitly adopted, and therefore to be matters about which there are facts to discover.

§8

CONCLUSION TO PART III

The general point of Part III was to evaluate the value of permutation arguments against the causal theory of reference, or views that take causality to play a major role in fixing reference. Their popularity, despite their unsoundness, is due to three sources.

First is the tendency to commit certain sorts of error. I've argued that this is operating in the later Quine, and, although disguised somewhat by semantic ascent, in Field (1975). The second source for the popularity of such arguments is an (often implicit) tendency to draw a distinction akin to a traditional observation–theory distinction. We find Davidson restricting our access to the data needed for evaluating reference to the context of interpretation, where the notion of reference operates only within the confines of a Tarskian truth-theory. Quine rules out the required causal resources by trimming them into blocky units of nerve stimulations, which are his *ersätze* for sense-data. Putnam, finally, restricts what he calls operational constraints on reference (those which contain whatever causal resources are available to speakers) to observations.

Here's a melodramatic moral one might draw: Quine killed the positivistic distinction between theory and observation with his holism, and buried it with his naturalism.[1] The ghost of this distinction haunts Quine and his successors to this very day, exacting retribution (as ghosts do) by poisoning its victims with (in this case) vague instrumentalist promptings.

Here's a more charitable moral, and one I prefer: reference really does transcend causal resources available to us; in some sense, the causal theory of reference, and approaches like it, are wrong-headed projects, and (this is

---

1 I don't mean to suggest that logical positivism had only one murderer, for I wouldn't dream of arguing with Sir Karl Popper (1974: 69) when he writes:

> Everybody knows nowadays that logical positivism is dead. But nobody seems to suspect that there may be a question to be asked here – the question "Who is responsible?" or, rather, the question "Who has done it?" ... I fear that I must admit responsibility. Yet I did not do it on purpose ...

the *third* source of the appeal of permutation arguments) every one of the above philosophers recognizes this, to one degree or another. Putnam, indeed, insists on describing reference as a normative notion. Something of this normativity has emerged in the possibility of φ-languages. Further details about the normativity of reference, and what it comes to, are given in Part IV.

Part IV

THE TRANSCENDENCE OF REFERENCE

If you're asked: "How do you know that it is a thought of such and such?" the thought that immediately comes to your mind is that of a shadow, a picture. You don't think of a causal relation. The kind of relation you think of is best expressed by "picture," "shadow," etc.

(Wittgenstein)

§ 1

INTRODUCTION TO PART IV

Contemporary debate over realism comes in numerous flavors. One important version of this debate focuses on reference, for many realists treat the referential relation as a natural one holding between the use of a term and the object it denotes. This relation is commonly taken as (1) causal (or supervenient on such), and (2) a real relation between speakers and the world. Part IV is concerned with referential realism in this sense; and I describe my referential realist as committed to *naive naturalism*.[1]

There are some disagreements about what naturalism comes to, so I should say a few words about this. Contrary to the views of many philosophers, the naive naturalist accepts the use of standard mathematics *as is* in empirical applications.[2] Notions from the special sciences are also naturalistically acceptable despite the absence of physicalistic reductions (Part I § 5). No more does naive naturalism require a definition of "reference" in purely physicalistic terms, or even in non-physical but naturalistic terms, for instance terms for sociological relations, such as causal chains between users of the word "Aristotle" and Aristotle himself (Part III § 6). The referential realist doesn't need a definition for "reference" at all. Although not all causal relations are referential, it doesn't follow that the referential realist *has* to mark out precisely *which* causal relations are properly referential ones. He thinks the term "reference" already does this.

Still, *naive naturalism* thus understood offers restrictions: a *natural* relation must be accessible to humans. That is, we already have vocabulary in place that describes tangible relations between humans and the world. There are those available via the senses: humans can see, hear, smell, feel, and taste things. There are also those social relations available by virtue of what humans say to each other, what they read, and so on. And there are still others I haven't mentioned (Part II § 5). Although it may not be possible to

---

1   "Naive," because I'll argue for a version of naturalism that denies one or more of the conditions the naive naturalist is subsequently characterized as committed to.

2   See Azzouni (1994) and (forthcoming).

describe necessary and sufficient conditions on these relations that determine when they support reference, any purported instance of a referential relation not simultaneously an instance of (or composed out of) some combination of these is ruled out.

This requirement has bite. For instance, any notion of reference relying on faster-than-light transmissions, or which requires computational exercises beyond real-time human capacities, is not acceptable.

A second principle the referential realist (as I understand him) upholds is *the principle of conservation*: there must be no violations of ordinary referential practice as ordinary speakers understand it. If, ordinarily speaking, certain terms refer to certain items, this should hold of the referential relation as reconstructed by the referential realist.[3]

I show that, given our assumptions about what we refer to, the principle of conservation is incompatible with naive naturalism. What this implies philosophically depends, however, on *how* they are incompatible. Permutational arguments try to show that the referential relation is soft in a peculiarly *global* way: all the terms in our conceptual scheme (or language) are on a par as far as their incapacity to grip the world referentially is concerned, and the theoretical and empirical components of our conceptual scheme are holistically mingled so that they cannot be distilled out. My arguments do not motivate anything so strong: although we *cannot* naturalistically underwrite our referential relations, we *can* measure how far those resources fall short. And, in general, the gap between the referential scope of a term and our resources for underwriting that scope varies widely from term to term, and changes over time.

At this point, it's natural to propose *revisionism*. If one can measure where our naturalistic resources fall short, why not rewrite referential claims so that they fall strictly within those resources? I show that this isn't viable because it results in a language we can't use.

Nevertheless, realists can draw some satisfaction from the position sketched: there's no denying naturalistic resources as far as they go (and they go pretty far). We really do refer to items that impinge on us (or us on them)

---

3 Being a methodological principle, there's slippage. If certain intuitions used to elicit how ordinary speakers take reference to work don't fit with what's otherwise an appealing and simple theory, we'll let the intuitions slide. However, an explanation for the presence of these intuitions is still required. A responsibly designed semantic–pragmatic distinction is sensitive to this requirement.

Honoring the principle of conservation as understood here does not require embedding intuitions into principles in the most direct way possible. Folk physics embodies intuitions and expectations about how objects move when impinged on by forces. Some of this may be hard-wired. Newton's laws of motion require treating these intuitions as recognitions of quite local phenomena due to the presence of frictional forces. Aristotelian physics treats folk physical intuitions as embodying physical principles of motion in a more direct way.

in measurable ways. *But* realists must also accept that such impinging does not underwrite everything we *say* about the world, and that there's no way around this.

Outline of Part IV

I first review the original formulation of the causal theory (or picture) of reference for natural-kind terms, and show how the projection problem naturally arises out of it.[4] I build my picture out of the failure to solve the projection problem, conceding to the referential irrealist what's required by that failure, and no more.

---

4 This is the problem of how to extend the reference of a natural-kind term beyond the samples initially linked to it. The issue originates in Goodman (1955), but its role in current debates over reference and realism is masked because contemporary causal theorists allow themselves more latitude (in particular, natural kinds and modals) than Goodman did, and because anti-realist critics use permutation arguments.

§2

TROUBLES FOR NAIVE
NATURALISM

The causal picture of reference for natural-kind terms goes something like this.[1] We start with a *sample* of something.[2] The sample may be of a kind of stuff, like gold, or of a biological kind – some things in a cage that don't mind reproducing with one another; the sample can be of events, instances of a certain collection of symptoms, or examples of a particular process. In each of these cases, we use what's in hand (the samples)[3] to refer to an entire kind: In the first two cases, we refer to a kind that contains the sample – in the latter two cases, we might refer to a natural event-kind containing the sample; but, more often, we refer to a kind that contains the *causes* of the events sampled.

It's obviously valuable for us to describe more of the world than just the samples in hand. In particular, with gold we'd like to back up our frenzied search for the stuff with a predicate that holds of everything that is the same as our sample. One might enumerate a short (or long) list of properties the sample has, and use those as necessary and sufficient conditions for the predicate. The problem is that this list may not be either necessary or sufficient for capturing everything (or even anything!) the *same* as the samples in hand. We could just *define* our natural-kind terms in terms of plausible lists of necessary and sufficient conditions – except for strong intuitive support that many such terms, in natural languages at least, refer to whatever is the same as the samples available, rather than merely to whatever has the same plausible list of properties affiliated with each term.

Even a description of the situation as one in which a natural-kind term is to refer to whatever is the same as the sample is not entirely accurate, as

---

1   See Kripke (1980), and Putnam (1970, 1975a), among others.
2   We could start with a theory that a sort of something exists, and samples could follow later. This doesn't affect what I'm about to say.
3   Samples "in hand" or "available" are usually instances of such things in the hands of experts, or, more generally, uncontested ways of recognizing or generating samples that experts have, or that are available to the general population. So "in hand" is relative to the linguistic division of labor.

samples can be more or less pure. What's actually being looked for is whatever it is about the sample that makes it operate as it does (have the properties it seems to have). Samples of water are invariably impure, but nevertheless it's not a kind of diluted mud we mean to refer to by the term "water." Water, even the impure stuff, can do certain things and seems to have certain properties, and we mean to refer to whatever is the source of *them*.[4]

That the sample is crucial to the causal picture of natural-kind terms may seem sufficient to anchor the natural-kind term securely in the natural world by natural means: the term refers to whatever is the same as *that*. But this is so only if we *successfully* project the predicate beyond the sample or its particular cause(s). We (collectively speaking) are only causally connected to whatever samples we (collectively) causally interact with – so it's not that there's a causal connection between us and all the *rest* of the stuff out there. If we're inclined to take relativistic physics at all seriously (and who isn't nowadays?), we can't claim we're causally connected to any of the gold outside our light cone, although certainly we take "gold" to refer to gold stuff *wherever* it turns up. Neither can we simply reiterate talk of sameness with a gesture towards the sample, for this is merely a sleight of hand if no explanation for the magic in such talk is offered.

Giving an answer that fails helps illuminate the problem. Suppose we say "same stuff" or "same kind" is anchored via *samples*, just as the term "gold" is, and that our term "gold" picks out *everything* gold whether we're causally connected to it or not, because our term "same stuff" picks out all the same-stuffs. But, look: first, there's a regress, as we haven't been told how we're supposed to use the predicate "same stuff" to project *itself* over all instances of *same-stuffs* beyond the samples (if any!) of same-stuffs we already have samples of.[5] And, second, talk of sameness is too abstract – too theoretical – for this. It isn't the case that all natural kinds are kinds in such a way that as soon as we've seen a few we know what to look for (nothing special marks out gold and tigers as natural kinds so that we can go on to see why lightning and static electricity belong together, too). If natural kinds are the

---

4 My exposition here is designed to *temporarily* bury a problem: although intuitively it seems clear that natural-kind terms *don't* have sufficient and necessary conditions associated with them, they aren't always used to pick out respectable natural kinds in the scientific sense *either*. This is fairly obvious for animal-kind terms, but it holds even of terms for "stuffs." One problem, as we saw in Part II § 3, is that in practice we're always faced with impure examples of stuffs, and there is something unprincipled or even outright conventional about which impure mixtures we assimilate to what *kinds*. See § 5 below, for further discussion of this.

5 Does it just do the operation on itself? Or does a higher-order notion of similarity do the job for it? And if it does do the job on itself, why couldn't the other predicates do this job for themselves too?

building-blocks of the sciences, then it's only with quite a bit of research that one gets a handle on where they begin or leave off.[6]

Putnam argues for this when he says that "the relation same$_L$ is a *theoretical* relation: whether something is or is not the same liquid as *this* may take an indeterminate amount of scientific investigation to determine."[7]

I've presented this worry, not in a dramatic, first-order, model-theoretic form as Putnam does later (1977), but rather in a way that naturally arises from more specific concerns with the causal theory of reference (as in Putnam 1975a). Nevertheless, even these more modest fears can be disputed. One can argue that crucial to there being a projection *problem* for natural-kind terms is a surfeit of candidates for the appropriate extensions of the terms in question. *And* this appearance of surfeit is due to falsely assuming that any grouping of items, any *set*, can function as the extension for a natural-kind term. But why assume that? Rather, perhaps, sets divide into two types: the *eligible* sets, whose members are all and only those things that share some natural property, and the remaining sets. And, only eligible sets can be extensions of natural-kind terms.[8]

What makes natural properties special, on this view, is their marking out joints and cleavages of the universe – the universe falls naturally into parts and those parts are precisely the eligible sets. Nevertheless, we can still wonder how *we* manage to make our terms pick out eligible sets. After all, any samples *we* possess of a purported kind consist in only small non-eligible chunks of the universe (because presumably, such samples are both incomplete, and contain bogus members). Surely we *could* be so unlucky that whatever (referential) tools we use to expand and winnow these samples leads to non-eligible extensions for our terms. The concern is not the sceptical one about the existence of such joints in the universe: we can cheerfully grant that the universe falls naturally into certain kinds and still wonder what *mechanism we use* to refer to just those kinds. If a naturalized story

---

6   Switching to talk of the causal mechanisms underlying the natural units the world falls neatly into won't help either: causal mechanisms are every bit as heterogeneous as natural kinds are. The problems with both idioms are on a par.

    By the way, it should be clear that attempts to use scientific theory to help fix reference to instances of something outside our causal reach face the same problems that arise with the use of "same stuff."

7   Putnam (1975a: 225). Well, the point isn't *just* that it's a theoretical relation. It's that (1) it's not a relation we grasp all the instances of by observation, and (2) it's quite heterogeneous in the sense that the properties we use to grasp instances of it are not the same ones as we shift from case to case.

8   The original version of this argument is due to Merrill (1980); it's taken up enthusiastically by Lewis (1983). Field (1981) also endorses the same kind of move. I omit the niceties of Lewis' account (grades of eligibility, causal descriptivism) because they don't affect the objection I'm about to make. It's worth mentioning the anticipation of this approach, with misgivings, in Quine (1969b, see especially p. 131).

compatible with our human powers is needed *at all* to explain reference, then surely the issue just raised must be dealt with in order to guarantee that an explanation in terms of eligible sets involves nothing occult.

Lewis disputes the legitimacy of *this* claim:

> We have no notion how to solve the problem of interpretation while regarding all properties as equally eligible to feature in content. For that would be to solve it without enough constraints. Only if we have an independent, objective distinction among properties, and we impose the presumption in favor of eligible content *a priori* as a constitutive constraint, does the problem of interpretation have any solution at all. If so, then any correct solution must automatically respect the presumption.
>
> (1983: 54–5)

In other words, we must *assume* as part of our theory of reference that only eligible sets are, well, *eligible* as extensions for natural-kind terms. This, sadly, is a *stipulation* disguised as an *a priori* constraint (as they so often are); there's nothing wrong with this move unless you are a *realist*. For a realist takes eligible sets to exist independently of us, and we cannot *stipulate* a connection to something independent of us. Referential irrealists uncomfortable with realism solving its problems with stolen irrealist tools (*a priori* constraints, or *this* a priori constraint, anyway) can take Putnam's 1981 way out, which is to claim that there is no solution to the problem of interpretation, and adopt a version of internal realism or linguistic idealism instead (where stipulation *can* happily flourish without needing an epistemic justification for its success).

As far as referential irrealists are concerned, there may be an impasse here, but those sympathetic to naive naturalism should regard Lewis' strategy as inappropriate for solving the projection problem because it relies on "presumption" to acquire what should be earned only by hard work and toil. If *our* causal powers enable us to fix reference, then a story should be told that turns on *our causal powers*. Lewis' story needs a Kantian-style completion which tells us not only that the world sorts into eligible sets, but also explains why *our* powers enable us to pick out those sets (e.g. that the world *we refer to* is a world of possible referential contact – not an irreferable thing-in-itself). Putting the matter this way exposes Lewis as yet another Putnam (as if there weren't enough to go around already) who avoids revealing his commitment to "internal realism" by refusing to tell us *why* the *a priori* constraint he offers works.

If our causal resources exhaust themselves before they finish underwriting reference the way the naive naturalist needs, what else is possible? Well, there are two other familiar approaches. We can supplement our causal resources modally; or, we can take our natural-kind terms to be fixed not by means of our *current* causal connections to the world but to such causal connections idealized via new procedures available in the Piercean limit of scientific development.

Here's how these moves work. Consider the methods (procedures) available at any particular time for recognizing gold. Anything is gold, we claim, provided that, should it be presented to us, our methods certify it as gold. So, if a particular lump of something lies *outside* our causal reach, it's gold if, *were it within the purview of our methods for testing for gold*, those methods would certify it as gold.

Let's say this suggestion works as far as it goes.[9] Our methods, however, are fallible and limited in another significant respect. Consider the ancient Greeks. Their methods did not certify the presence of gold in a bowl of seawater even when it *was* within their purview. What about a possible situation where the minute amount of gold in question is conveniently located within a larger lump of the same substance? Unfortunately the same positive result would be yielded no matter what identically minute amounts of a substance were substituted for the gold.

We therefore cannot rest with the particular procedures associated with a term that are available at *any* time[10] – new ones are developed, old ones refined, and all such developments are taken to pick out hitherto hard to locate items which the term is already taken to refer to. We can't extend the modal suggestion by itself to handle this problem, unless the natural kind itself is used to fix reference pretty much as Lewis urges. For how else are we to describe possible situations, in which we're using new additional procedures to recognize something, and guarantee that the additional procedures pick out the same things that our earlier set picks out (without simply stipulating it)?

This, therefore, makes it natural to combine the modal approach (which is still needed to handle those instances of anything that lie outside our light cone) with the Piercean limit approach to procedures.

Some may be puzzled. Are we still within the confines of naive naturalism? What follows is an elaboration of this worry.

Suppose reference is not underwritten by causal relations; suppose, indeed, that reference is not a naturalizable (or natural) relation *at all*. What would the implications be? Let's leave aside, for the moment, worries about appropriate scientific methodology,[11] and get a feel for what's really involved here.

---

9   One can worry that it doesn't work because it's not clear that the sort of possibility required, *physical* possibility, can successfully underwrite the subjunctives needed. Let's waive this fear for the moment; I say something about it presently.

10   Even the most arrogant apologist of science must concede that this includes the present time. Our methods for detecting many substances are limited and specialized, and we regularly develop new ones (Part I § 7).

11   I'm alluding to an argument found in Field (1972). See § 5.

Certainly one implication of the suggestion that reference isn't under-written by causal relations is this: we can't use underlying scientific laws as constraints on data about reference. Consider an analogy: if biological phenomena were not constrained by physical laws, we could not apply phys-ical law to biological items even when tractability considerations made it otherwise possible. So, given a worst-case scenario, we could not apply aero-dynamical insights to help explain the flight of birds, an admittedly bizarre situation.

Would something similar happen with reference? Well, it doesn't seem to. Instead of figuring out exactly how causal chains between us and the objects we refer to can be described, so that we see how referential relations to objects outside our light cone *are* causally possible (e.g. invoking non-locality), we instead invoke currently *nonexistent* procedures for recognizing instances of kinds, and possible but *unrealized* testing situations, to under-write a referential relation that we take to be *already* in place.

Consider birds again. Suppose we don't understand how, aerodynami-cally speaking, a bird (of a particular species) manages to fly. What would we do? There are four – not equally palatable – possibilities. The first is to search for factors overlooked (or undiscovered) that we had failed to take into account. Perhaps the bird's feathers are unusually shaped, and this enables it to gain the needed additional lift we had trouble explaining. Or, perhaps, there are hitherto undetected wind currents the bird uses. Or perhaps ...

If this approach failed, even in principle, to show how laws of aerody-namics can be applied to the flight patterns of this sort of bird, three possibilities remain: we'd give up for now (this happens!); or we'd entertain the suggestion that the laws of aerodynamics as currently formulated are wrong; or we'd consider the possibility that physicalism is wrong – that there are emergent biological phenomena, and, surprisingly, that the ability of these birds to fly is among them.

As I said, by no means are these possibilities equally appealing, but contrast them with an option we *wouldn't* consider, even for a moment. We would not sketch out alternative abilities for the bird (bigger wingspans, lighter bones), or alternative scenarios (weaker gravitational force, shorter distances to travel) and, in describing how birds could successfully fly under *those* circumstances, convince ourselves we'd explained, in a way compatible with physicalism, how birds, *our* birds, can fly.

Now perhaps this analogy is mistaken. Flying is something birds do (birds fly to Mississippi). And although reference can be seen as something people do (people refer to Mississippi), it's better, so one might argue, to think of reference as a dispositional property: "referring to A" is something like "*being able to be* causally in contact with A (in the right way)."

Putting things this way justifies taking reference to be fixed, not by causal relations alone, but by such augmented with counterfactual tools, and the procedures available in the Peircean limit. Notice, however, two interesting points. First, we were driven to consider modal supplementation and idealized procedures *not* because it was obvious that when we refer we're doing something dispositional (in a way that it's obvious the *ability* to fly is dispositional), *but* because reference couldn't be underwritten by actual causal relations *alone*. So there's still a methodological disanalogy with birds; it would be surprising if the reason we can't apply aerodynamical laws directly to the flight of birds is because flying is a disposition and not an action of birds. Reference, too, pretheoretically considered, doesn't seem to be a dispositional property: it seems to be something we *actually* do, not something we *can* do. Having made this point, I'll set it aside in what follows.[12]

The second point is this. If we're really to honor naive naturalism, it won't do to merely invoke modals and future procedures. For one doesn't explain dispositions by lipsyncing such apparatus irresponsibly: one makes sure there's a mechanism the disposition is rooted in. I'm not objecting to taking dispositions, and consequently the counterfactuals supported on them, as surd or brute phenomena not to be explained in terms of a mechanism, by the way; I'm objecting only that if one does so, one should not simultaneously claim one is honoring the strictures of naturalism.[13]

Well, given this last point, *is* the suggestion (of underwriting reference in terms of causality + modal supplementation + refined procedures available in the Peircean limit) naturalistically acceptable? Despite the last footnote's warnings, modal supplementation is fine because the procedures themselves provide the mechanisms via which the dispositions needed can be described; but there's a problem with the Peircean limit,[14] for notice that what's

---

12 The status of a relation, pretheoretically considered, is not sacrosanct. But the disanalogy indicates something methodologically suspicious. Naive naturalism is supposed to constrain our theorizing about reference. The way modality and idealized procedures are brought into the picture hints that our pretheoretical views about reference constrain naive naturalism instead. I worry more about this shortly.

13 It's ominous that modality arises so often in the context of philosophical programs which attempt to reduce something to something else: theoretical entities to observational ones, material objects to phenomenalistic feels, and yet modality is never a tool that arises when *scientists* are concerned with the reduction of entities in one special science (e.g. chemical kinds) to those in another science.

I'm not suggesting that modality and dispositions aren't scientifically respectable, for dispositions can be law-like. That quarks have the properties they do explains a lot. Nevertheless, such properties are dispositional properties (something with such and such a property acts in such and such ways under such and such circumstances), and one for which at present we have no explanation in terms of underlying mechanisms. Invoking dispositions to underwrite laws when one hasn't an explanation of such dispositions in terms of underlying mechanisms is one thing; using counterfactuals to make ontological constraints appetizing is another.

14 A number of philosophers think it unlikely there *is* a Peircean limit for scientific theoretic

required is not merely a completed theoretical science, but a completed *applied* science. This presumes idealized science can standardize all procedures, enabling us to specify, ahead of time, under what conditions any device used in a procedure will malfunction; and it also presumes that we'll expand the scope of our procedures for any term so that (collectively) they're both bivalent and consistent.[15]

Nevertheless, although the proud user of idealized science need not be omniscient, it's no exaggeration to call her "omnipotent." For although a completed theoretical science is not required (it's not necessary that our theory of how we pick out objects be right), still it's very unlikely that all possible applications of science could be completed and standardized (so that we always know what to do), and that these standardized practices would be sufficient and necessary to pick out the extensions of our terms. For an implication of this is that the application of the sciences to devices designed to make causal connections to objects, and to gather evidence about them, could be done in such a way that total conditions under which such devices suffer failure or enjoy success are given explicitly.

Theoretical deductivism makes a completed applied science seem plausible (at least in principle). For on that view we *derive* applications from scientific theory. Theoretical deductivism is wrong (Part I § 3), and so the suggestion posed here is epistemically incoherent. Recall that the auxiliary apparatus of science, in general, is so complex that the design and operation of devices, for the most part, is necessarily sloppy and empirical. *Ad hoc* adjustments on the basis of gross regularities are always necessary. Recall that, in general, even if a set of scientific laws is mathematically tractable, this doesn't imply that the *application* of such laws to most situations must be mathematically tractable. On the contrary. As a result, scientific application requires approximating models, mathematical short-cuts, and other tricks which, in principle, are not standardizable. Speaking of idealized applied science with standardized application conditions for every procedure is pretending that sense can be made of the idea that this sort of thing (mathematical short-cuts, idealizations of all sorts, the use of gross regularities) could become superfluous.

If what I've claimed so far is right, we can understand why any attempt to broaden or supplement our notion of *causal mechanism* so that

---

development. See, among others, Putnam (1978a: 36); Rorty (1986: 129–31); Williams (1980: 269). I use the adjective "theoretic" because the suggestion is attacked by these philosophers in the context of evaluating the project of defining "true" as, say, "what will be verified under ideal circumstances of inquiry." Worries about the Piercean limit are different when the issue has to do with reference, as I show momentarily.

15 Both consistency and bivalence could be given up by changing our logic. The considerations just ahead, however, show that these radical moves would not save the suggestion anyway.

it underwrites our referential practices is doomed: It requires making sense of the project of second-guessing applications of scientific development.[16]

Something else follows from these considerations. As I've already shown, we cannot simply underwrite reference in terms of actual causal connections to items we refer to; it must be done in terms of *possible* causal connections to items we refer to. And fair sense can be made of this "possible" provided we fix the set of procedures needed for any particular term. For the underlying mechanism relevant to the dispositions invoked here are found, fairly unproblematically, in the mechanisms of the procedures themselves.

But since the set of procedures causally connecting us to instances of a kind have to be open-ended, the dispositions must be supported not by the procedures themselves but *by us as a community of researchers*; and it seems unlikely that there are mechanisms in us or in our institutions that can be used to do this. Although, in hindsight, we can often see why certain scientific developments made it likely that a certain mechanism would be designed, it's too much of a stretch to suggest we can justify dispositional talk in the case of the ancient Greeks that explains why it's not naturalistically irresponsible to describe *their term for gold* as referring to the minute gold in seawater. In short, an explanation of reference in terms of causal connections forged by new procedures is *occult*: the way applied science develops new procedures does not turn on the existence of a mechanism of a certain sort, either in us or in the institution of science.

I turn now to an aspect of this project that makes its motivation puzzling. Suppose I'm right about the future of applied science; or, if you prefer, suppose I'm *wrong*. What difference does it make? In what way does the nonexistence of the Piercean limit (for applied science) affect what we do or how we talk; in what way does a failure of the modal approach to reference affect our referential practices (affect, if you will, *any* of our current practices in science and in ordinary life)? The answer is *in no way at all*. Regardless of whether there's a Piercean limit for applied science waiting out there at the end of time, our referential practices are exactly the same.

This is a deep and unnoticed methodological point; although it can look trivial if you observe that the concern all along has been with evaluating a *theory* of reference – and surely a failure to develop a successful *theory* affects our referential practices as much as a failure to explain how birds fly affects *their* practices.

There's a sense in which this is right. But, more interestingly, there's a sense in which it's not. Imagine two possible worlds where the inhabitants

---

16 One may hope certain terms, "cow," for example, are exempt from this argument because there is a relatively limited number of procedures science could design for causally interacting with cows. Alas, this is a version of the move quashed in Part II § 3, of marking out a class of terms with "ordinary" criteria of application; it's no less hopeless simply because we've got another motive for wanting it to be true.

have identical science and identical referential practices for the next 200 years or so. Imagine, in addition, the brains of (some) of the people in each world differ, so that in one world a Peircean limit for applied science exists, and in the other world it doesn't. How might this happen? Well, suppose the Piercean limit requires the development of a kind of very sophisticated mathematics that makes certain scientific applications trivial; and suppose that although it's not difficult to *learn* the mathematics from someone who already knows it, it's impossible to *invent* unless one can generate a certain sort of very weird mental imagery. In one world folks with such a capacity are sprinkled among the population; not so in the other world.

Here's the really odd fact: *despite currently identical referential practices, we must tell very different stories about reference in either case.* For only in one world does the Peircean limit exist.

Perhaps this comes as no surprise. For imagine two possible worlds, one where the world is pretty much as we think it is, and the other in which, alas, we (or *you*, anyway) are in the grasp of Descartes' evil demon. Here, too, two rather different stories have to be told about what we (you) refer to; change the metaphysics and you change whatever depends on that metaphysics.

But the two worlds *I* described don't differ metaphysically, and the people themselves don't differ very much either, not enough, one would think, to expect different descriptions of how reference works among them. We recognize one group as able to implement applications that the other group can't – about things they *both* refer to. That's how we would ordinarily describe the situation. In relying on the Piercean limit for applied science, that is, we seem to go quite beyond the data to be explained.[17]

These considerations motivate telling a different story about how our referential practices operate, and how causality is involved. The story I'm partial to is as naturalistic as the one attempted here, that is, it also makes clear how creatures such as ourselves can have such practices. But it won't show that reference is underwritten by causality, for instance that any referential relation is, in some sense, solely constituted of causal ones.

Here's the state of play. I showed in Part III § 7, by use of φ-languages, that the referential relation could fail (pretty much) to be constituted out of causal relations. I have just shown that such a failure holds, in some sense, of our own language as well. I've taken a bucket or so of cold water, and thrown it on certain suggestions for supplementing our causal resources by including additional resources we'll have at some indeterminately future time, or could have (if circumstances were a little different), and underwriting reference in terms of *those*.

---

17 Suppose *both* groups extinguish themselves (because of atomic weapons) before the deviations in their abilities to apply science emerge in deviations in their history.

Although I've not used the jargon, it should be clear that my objections apply to attempts to naturalize the referential relations that are generous in resources. For example, reading the referential relation as a second-order physicalistic relation (that is, as a functional relation with many different physical instantiations), and requiring a law-like connection between term used and object denoted, won't help. It's simply not plausible to think new procedures that pick out new instances of something can be subsumed under the same functional relation the old ones belonged to. Consider our methods of recognizing the presence of gold by our senses, and the method of recognizing its presence in solution by some chemical process or other (adding a chemical to a liquid which turns it green, say). These are neither subsumable under the same law-like connection between our use of the term "gold" and its instances, nor under the same functional relations.

In § 3 I examine some widely held intuitions about reference. Then I consider the revisionist move of jettisoning our notion of reference and substituting one that is more compatible with the aims of naive naturalism.

§3

THE ELUSIVITY OF
REFERENCE

The reader might suspect that part of what's behind the just-rehearsed failure to underwrite reference by causal relations (plus other naturalistically acceptable resources) is something similar to the elusivity of truth (Part II § 6). The suspicion is correct: Reference cannot be *defined* in terms of underlying causal relations, nor can causal procedures of any sort be used to augment or restrict our notion of reference, just as truth cannot be characterized in terms of whatever truth-gathering methods are available at a given time. There are versions of Putnam modals about reference that we intuitively find true: "There might be instances of gold we could never causally interact with no matter how much we perfect our procedures" or (for any *p*) "what causal process *p* picks out might not be (just) gold," are examples. These can be used to show points about reference analogous to points I've made earlier (using Putnam modals) about truth.

Unlike referential irrealists, I've not wanted to draw the conclusion that causality has nothing to do with reference, any more than I've taken the elusivity of truth to show that our current truth-gathering procedures have nothing to do with truth. The arguments offered in Part III § 6, that causality (of various sorts) is robust, allows a study of exactly what role causality plays in our referential practices.

But apart from the somewhat distinct issue of the relation of causal processes to reference, I'd like to focus attention on a refinement of the elusivity of reference. The points to come have not been clearly seen in the literature because they're classified "metaphysically," as it were. But they actually amount to a linguistic distinction between *criterion-immanent* and *criterion-transcendent* terms. As I show, terms like "truth" and "refers" are criterion-transcendent, whereas terms like "warranted assertibility" are criterion-immanent.

A methodological point. I use a number of well-known thought experiments from the literature – most due to Putnam and Kripke – to motivate the distinction. Both these philosophers, however, ply these thought experiments for ambitious purposes, to show that sentences such as "cats are mammals," are not analytic, or to show that certain views about how the references of natural-kind terms are fixed are wrong. My purpose is humble: to show that the criteria for criterion-transcendent terms operate in a different way than they do with criterion-immanent terms.

In order to make this distinction I start with a prior distinction between criteria and procedures. The latter, recall, are methods of forging or exploiting causal connections between us and the items referred to. Among these are *recognition procedures*, methods by which we identify things our terms refer to. Some ways of recognizing things, however, involve actually *making* them. The standard high-school procedure of generating water from hydrogen and oxygen is a method which by itself can be used to recognize that it is *water* droplets that have been produced. Call the methods for making things *production procedures*, and notice that not all recognition procedures are production procedures, although, in principle anyway, any production procedure can be used to recognize what it results in.

Criteria are different: We often describe kinds in ways which themselves are not descriptions of procedures for picking out instances of such kinds, but are at most descriptions of the kinds that enable us (in principle) to generate procedures for recognizing instances of the kinds.

Take natural-kind terms first.[1] A current description of (pure) gold requires that it be (solely) composed of atoms with atomic number 79. This doesn't by a long shot describe a *procedure* for recognizing gold (try counting the protons in an atom), although that gold has atomic number 79 can be (and has been) used to design such procedures. Call this a *criterion* for gold, and describe the *criteria for t* as the set of these things we take to hold of *t*.

Criteria for natural-kind terms change: at one time, before the emergence of sophisticated science, being heavy, yellow, inert, metallic, highly ductile, and so on, were among the criteria for gold. This list can be (pretty closely) identified with procedures (available at the time) for recognizing gold, although this doesn't have to be the case. Procedures can't be *identified* with criteria because procedures are tests to see, in these cases, whether the criteria *hold*, and, as always, such tests may seem to show that the criteria hold or don't hold, contrary to fact. Still, the criteria themselves when it comes to natural-kind terms, even and above the epistemic doubts possible that a procedure has shown they apply, are best thought of as provisional "satisfaction-conditions" of such terms.[2]

---

1 In using the terminology "natural-kind term" I'm not suggesting the kinds designated are "kinds" in the (one or another) scientific sense, for they often are not. "Mouse" holds of both mammalian mice and marsupial mice, and "porcupine" holds of genotypically quite distinct animals.

2 I don't claim that each natural-kind term comes with its own special criteria which mark out what it is apart from anything else. Criteria for natural-kind terms are dependent on either ordinary or scientific views about what something is; we can have blocks of theory in which an indeterminate number of sentences with a particular term *A, and perhaps other terms, too*, appear, and which are taken to describe the properties, relations, etc., of what *A* refers to (in conjunction with the properties, relations, etc. of what these other terms refer to).

Now consider the criteria for terms of a different sort: "refrigerator," "notarized document," "hammer," etc. The criterion for a refrigerator is something like this: a device designed to house food and drink while keeping them cold. The criterion for a toy is something like this: any small object designed to be played with by children.[3]

A striking difference between the criteria for natural-kind terms and for the terms I'm considering now is that we allow the criteria for each sort of term to change in different ways. Recall the thought experiments (e.g. Kripke (1980); Putnam (1975a)) that show the ways we're willing to change criteria for natural-kind terms. We might learn that the atmosphere interferes with visual perceptions so that many objects we think yellow (such as lemons) are actually blue. Or biologists might learn that a certain sort of animal we take to be a kind of rabbit is actually a kind of insect. Or, more drastically, we could discover that lemons are not fruits at all, but a hitherto unknown stage in the life-cycle of (certain) squirrels.[4] With only a little imagination we can tell stories where things in the natural world are discovered not to be as we currently take them; such stories intuitively require massive reclassification of natural categories, with concomitant changes in their criteria.

Matters, however, are not the same with certain artificial-kind terms: *their* criteria seem immunized against such stories.[5] One cannot discover toys are actually bombs or pencils by discovering that part of the *criterion* for a toy *should be* that it explodes, or that it can be used to write on paper. One *could* discover that every object considered a toy to date is actually a disguised bomb (or pencil),[6] but these thought experiments do not give the same

---

3 Like the criteria for kind terms in the special sciences, these are open to problems of vagueness. Other objects, not initially designed as toys or refrigerators, could be drafted for such use (cold caves or streams as refrigerators, clothes-pins for toys, etc.). Were these things used as refrigerators or toys for a long time, and were they so successful at satisfying the purposes refrigerators and toys are normally put to, that clones of them were subsequently designed as refrigerators or toys, then the original objects so drafted (and still in use, let's say) would be called "refrigerators" or "toys," and doing so would be correct usage despite the apparent violations of the criteria for these terms. In trying to adjust the criteria given to handle such penumbral cases, one faces relatively irresolvable borderline questions such as, how long could such a thing be used as a toy before it really is a toy? There are issues here, but I won't pause to consider them now. See, however, the discussion of vagueness in §6.

4 Many of these examples are implausible given the vast amount of information we currently have about the animals and plants in question. I agree, but that's not the point. For actual examples, one should turn to the history of science: e.g. shifts in the notions of *element* or *fire*.

5 With one qualification: one may tell stories where an artificial kind turns out to be a natural kind, or *vice versa*. See Putnam (1975a). I touch on these examples shortly.

6 Whether the upshot would then be that no toys have actually ever been manufactured, or, rather, that every toy manufactured until now is *also* a bomb, turns on the story told, although with some stories I can imagine intuitions are indeterminate.

radical results about criteria that are possible with natural-kind terms. Similar remarks may be made about "refrigerator," "notarized document," etc.[7]

Thought experiments can also be designed which allow the shifting of a term's status from natural to artificial, or the other way around. Some of Putnam's most famous examples are like this: the discovery that all the creatures we call, and have ever called, "cats" are actually robots secretly controlled by Martians. In such a case we will conclude that cats are robots. We will, that is, change not only the criterion for "cat," but its status as well. Similarly (this is Albritton's example), suppose it's true that pencils are organisms of a certain sort (we uncover an insidious corporate conspiracy to portray pencils as manufactured objects); then "pencil" will undergo a criterion shift the way a natural-kind term does, but only because it has simultaneously shifted status *to* a natural-kind term.[8]

Artificial-kind terms such as "car," "notebook," "screwdriver," etc. as well as legal terms, such as "marriage," "notarized document," "legal tender," etc., are rather like "refrigerator" and "toy," in that they have criteria immune to Putnam-style thought experiments,[9] and this in contrast to natural-kind terms such as "lemon," and "tiger." The latter are *criterion-transcendent* terms, and the former are *criterion-immanent* terms.[10]

The failure of Putnam-style thought experiments to allow shifts in criteria for criterion-immanent terms does not mean such shifts are impossible. On

---

7   Burge-style examples can arise with either natural- or artificial-kind terms. (See Burge (1979).) Due to my own ignorance about toys, *I* might discover that all toys are actually bombs, that in fact to explode *is* part of the criterion for a toy. The distinction I'm marking out now abstracts away from anything due to the linguistic division of labor. By the claim that we could not discover that toys are actually bombs, understand implicitly the claim that we (collectively speaking) *could not* discover that toys are actually bombs, and similarly with the examples that follow. Burge-style examples exploit differences between the public, or official, criteria for terms and the individual's imperfect understanding of those official criteria. But I am concerned with differences in how official criteria can change.

8   It's not clear to me that it's always possible for artificial-kind terms to shift in status in this way. "Pencil," for example, might instead split into two homonyms: one applied to the organisms, and the other to the objects manufactured in their stead. Also, suppose it turns out that *everything* we call a toy (and have ever called a toy) is actually a cocooned Martian. I don't think anyone would draw the conclusion, upon discovering this, that toys just *are* cocooned Martians. Rather, we'd say there are no toys around at all (despite the well-intentioned attempts to construct such on the part of toy manufacturers).

9   Provided they keep their non-natural-kind status. From here on I'll be concerned only with how changes in criteria and/or procedures are allowed, provided the terms in question do not switch categories from criterion-transcendent to criterion-immanent, or *vice versa*.

10 Why not "natural-kind terms" and "artificial-kind terms"? Because the sets of terms I'm marking out go beyond what are normally picked out by these phrases, and because these phrases imply something I explicitly reject, things I show in a moment.

the contrary: but we must get at them by a rather different sort of thought experiment. At present screwdrivers are used for a certain purpose, and are designed for that purpose. But suppose some other purpose arises that screwdrivers can also be successfully used for (unlocking "glibs," a type of bolt, let's say). In time, the design of screwdrivers could evolve slightly so that the tool is optimized for both purposes, and, simultaneously, the criterion for a screwdriver could come to include not just screwing screws but unlocking glibs. Precisely this sort of evolution goes on, as a matter of course, with legal concepts.

Notice the difference between this sort of change in the criteria for criterion-immanent terms, and the sort of change we contemplate when constructing one or another thought experiment for a natural-kind term. In the latter case, the thought experiments show the criteria for a criterion-transcendent term can turn out to *fail* to fit many or all of the instances of the kind in question, and in fact *never* to have fitted any of the items the term applies to. In the former case this is impossible.

There are intuitive differences between these sorts of terms in respect not only of criteria but of procedures. Criterion-transcendent terms such as "tiger" and "lemon" admit of procedures that are not user-relative (except in the matter of skill). If I use a certain procedure on something and get one answer, you use it and get another, we think something has gone wrong, either with the procedure or with one or both applications of it. "Going wrong," includes cases where, in applying the procedure, one or both of us has wandered into the vague penumbra where the procedure is "unreliable."

Terms such as "toy" or "refrigerator" also have procedures associated with them, just as natural-kind terms do. An example of a recognition procedure for a toy is finding the appropriate toy-manufacturer label (Hasbro) on it, or finding it for sale in F.A.O. Schwartz, or perhaps seeing that it's small, brightly colored, a (cheap) miniature imitation of something children like (such as a car, a boat), etc.

But any of the procedures for these artificial-kind terms can become out-and-out *inappropriate* if things shift enough (culturally speaking) in one locale; and procedures that are inappropriate means of recognizing something as a toy (say) in this country (or district) might be quite appropriate elsewhere. The same physical object that is a toy here could fail to be a toy there. Both production procedures and recognition procedures for such terms obey "locality constraints," or relativity to time and place.

Another example. At any one time there can be legal and cultural production procedures associated with the term "marriage." These can change over place or time, so that what's called for in being married at one time and place is not what's called for at another; and despite our agreement that the criterion for marriage (when described abstractly enough) is unchanged.

The dichotomy being explored here goes beyond the examples considered thus far: criterion-immanent terms include more than the nomenclature for artificial kinds, tools, or certain legal states. Consider "warranted assertibility," a normative term, and apparently one in a different logical class from the previous items. As I observed in Part II, we can ask, "*should* a claim gotten by such and such a method be warrantedly assertible?" and we may decide it should not, that the method (perhaps one we've used for a long time) is not good.

Nevertheless, the claims warrantedly assertible at one time need not be warrantedly assertible at another time, although shifts in what is warrantedly assertible (and in the methods for establishing warrantedly assertible claims) do not change in a way that affects the notion of warranted assertibility itself (for example, part of the criteria for the notion is that methods which produce warrantedly assertible statements *strive* for truth).

These facts show that "warranted assertibility" is criterion-immanent. "True," by contrast, is criterion-transcendent, and so it shares significant properties with natural-kind terms, as well as terms for artificial products such as "plastic," or "nylon," and certain broader philosophical notions such as "knows" and "refers."

Two important observations. It's the criterion-transcendence of the classical notion of truth, I surmise, that makes philosophers, such as Dummett, who in some broad way take meaning to be use, so worried about our capacity to understand it. Similarly, that many notions of ours are criterion-transcendent is something those influenced by Wittgenstein have trouble accounting for. In fairness to them, if "satisfaction-conditions" is understood in a criterion-transcendent way (as one could easily want to, since *truth* is so closely connected to *satisfaction*), then it's hard to see how such things can be part of the semantic competence of a speaker and yet be "psychologically real" – something speakers *really* grasp. This motivates replacing "satisfaction-conditions" with something else, "assertibility-conditions," say, because assertibility conditions have a better chance at being psychologically real since they're linked to the actual procedures used by the speaker (or her community). But this can't be the *whole* story about a speaker's understanding of (and practice with) criterion-transcendent terms, for it doesn't include how the criteria and procedures for these terms change.

Second, it might seem that there is a serious disanalogy between *truth* and the other criterion-transcendent notions: namely, that it admits of a criterion which doesn't change the way criteria for criteria-transcendent terms do (i.e. Tarski-style definition, or axiomatization). However, the Tarski adequacy condition yields the sentences of (T),[11] and *these* connect truth to criterion-transcendent notions via the truth-values of the sentences

---

11　See Part II § 7.

these notions appear in. Truth thus acts like a criterion-transcendent term in respect to the three intuitive differences forthcoming between transcendent and immanent terms, and it does so because it has inherited these properties from its logical role with respect to sentences.[12]

What are the intuitive differences found thus far between the immanent and the transcendent? First, the procedures associated with criterion-immanent terms, but not those associated with criterion-transcendent terms, can be relative to communities, individuals, or whatever, without our having to adjudicate which groups and individuals are right or wrong. This is why there can be warranted assertibility for me and warranted assertibility for you (that is, the epistemic practices warranted for me need not be the same as the epistemic practices warranted for you), and why what I do to get a document notarized here can differ from what you must do to get a document notarized there. Tigers, by fierce contrast, are taken to be the same kind of thing for both of us – and similarly for truth.

The second intuitive difference between criterion-transcendent terms and criterion-immanent terms is linked to differences having to do with their criteria. Criterion-transcendent terms are ones for which we can imagine possible circumstances in which we drop the criteria in use for other criteria – and find ourselves asserting that the old criteria *never held* of either some or all of what we previously took them to hold of. Criterion-immanent terms, on the other hand, are not ones for which such possible situations exist. Rather, we can see how criteria might be replaced, so that they might subsequently fail to hold of some or all of the things they once held of, but we won't claim that the earlier criteria *never* held of all or some of the things we once took them to hold of.[13]

Similar intuitions may be elicited about procedures, but matters are a little delicate. Procedures for recognizing or making a kind of thing K can be faulted either because the criteria heretofore applying to K have been discovered to be wrong, or because the procedures are discovered not to satisfy the criteria for K in one way or another. The former sort of failing is not

12 Of course, truth is also connected to criterion-immanent terms by (T), but criterion-transcendent notions stamp truth as their own because the characteristic shifts the latter undergo translate into similar shifts in the truth-values of the sentences they appear in; and this affects *truth* even if the truth-values of *other* sentences are not so affected.

13 Mathematical-kind terms do not fit nicely into the transcendent–immanent dichotomy: they don't belong with criterion-transcendent terms, for when it comes to reference we tend to respect the criteria used at a given time for a mathematical-kind term (so that d'Alembert's notion of a function is not taken to include everywhere discontinuous functions in its extension). But they don't belong with the criterion-immanent terms either, for we also allow changes in criteria for such terms to be taken as correcting mistakes in what we took such terms to refer to. Mathematical terms are a special case, and must be explained via the special practices they are involved with. For details, see Azzouni (1994: Part I §§ 5–6). One significant factor is that there are no procedures for them.

possible for procedures of criterion-immanent terms, but the latter sort of failing applies to all procedures. We can discover that a certain method of designing devices that cool objects just doesn't work (the resulting contraption doesn't cool *anything*), but we can't discover that this method doesn't yield refrigerators, even though the resulting objects *do* satisfy the current criterion for refrigerators (contrast this with, say, methods of breeding tigers: we can learn that our criteria for tigers are wrong, and consequently that those things running around the farm aren't *tigers* at all).

There is yet a third intuitive difference between the two sorts of term, although it's been somewhat neglected. Take criterion-immanent terms first. If we change either the criteria or the procedures for a criterion-immanent term, the change is either, as I'll put it, *nonretroactive*, or *weakly retroactive*. If a new kind of refrigerator is introduced by a company (and it comes to dominate the market), no one thinks that discontinued models are not refrigerators. So, too, when new toys are invented; old toys may be set aside, but they don't cease to be toys.[14] These are illustrations of nonretroactive terms; the change in criteria or procedures for these terms has no effect on items that fit the previously acceptable criteria and procedures for the terms: they're still instances.

Consider, on the other hand, "legal tender," or "notarized document," where, although previous instances of legal tender or notarized documents are still *previous* instances of legal tender or notarized documents, they are unacceptable as *current* instances of legal tender or notarized documents. Similarly, we regard medieval thinkers (or many of them, anyway) as warranted in asserting the flatness of the earth, even though *we're* no longer so warranted. These are weakly retroactive criterion-immanent terms. A change in criteria or procedures changes *current* instances of the term, though not what were taken (correctly) to be instances of the term in the past.[15]

By sharp contrast, if we change the criteria or procedures for a criterion-transcendent term, the change is *strongly retroactive*: instances that don't fit the new criteria or which aren't picked out by any members of the new set of procedures are not taken to belong, *nor to have ever belonged*, to the extension of the term; so, too, items that fit the new criteria and/or procedures, but not the old ones, are taken to belong and to *always have belonged*, to the term's extension. If we change what's taken to be gold, instances that don't fit our new criteria are dropped from the sample collection, and others may be added. So, too, if we change what we take to be true.

---

14 Even when they fall into the hands of the avaricious adult collector they are still *toys*.

15 Nasty political moves are always possible. One can pass a law and apply it *ex post facto*, decide, on the basis of there being new requirements for a marriage-license that no one was ever married in the past. But this is the unusual case with criterion-immanent terms.

The third intuitive difference complements the second. Notice, however, that the second intuitive difference arose from contrasting Putnam-style thought experiments with thought experiments available for criterion-immanent terms. Although such thought experiments focus us on how often we keep samples of something, when we discover that previous criteria did not hold of them (and so change criteria to match our new view of these samples), they cause us to overlook the equally common winnowing and outright exclusion of items previously taken to belong to a kind in light of changed criteria. If certain reptiles are similar to tigers in certain respects, earlier categorizations which placed these tigers and reptiles together can be rejected in light of more refined criteria.[16] What's important is that we change the criteria for, and adjust the extensions of, criterion-transcendent and criterion-immanent terms, in different ways.

The initial description of the distinction between criterion-transcendent terms and criterion-immanent terms might give the impression that the distinction is a metaphysical one between terms taken to pick out something in the universe already antecedently there, and terms marking out an artificial distinction related to our own interests. Tigers, and what makes them alike, are the work of God: Adam has only a nomenclatural job regarding them. Chairs, on the other hand, are comfortable joints in the universe, which Adam first *carves out* for himself and *then* subsequently names.

This won't work. "Plastic" is criterion-transcendent, as I've already noted, and so are terms for genetically engineered biological-kinds. Even certain *legal* terms are criterion-transcendent: "murder," for example. On the other hand, as with the previously mentioned "mouse" and "porcupine," there is often no good scientific reason to treat a natural-kind term as picking out a genuine kind, and therefore as picking out a joint in the universe. Furthermore, this can be widely known and yet have no impact on the classification of a term as criterion-transcendent.

We need, therefore, some other way to explain the intuitive differences between criterion-transcendent and criterion-immanent terms. Here's a promissory note. Eventually (in § 5) I'll explain the value of introducing criterion-transcendent terms in a way that turns on their practical indispensability, and in so doing I'll explain their presence in our language without making strong metaphysical assumptions about differences between artificial kinds and natural kinds.

I'll make some final remarks about "refers" before closing out this section. It's a short step from the existence of criterion-transcendent terms to the fact that our notion of reference itself must be criterion-transcendent (this follows by semantic ascent as with truth: the criteria for "refers," and its status, are

---

16 The repudiation of whales as fish is a classic example.

closely connected to the criteria for terms such as "tiger," and their status). The criterion-transcendence of "refers" is also seen by observing that "refers" is intuitively treated much the way "tiger" is, rather than as "notarized document" or "refrigerator" are. But there's something else to observe about reference: It is, as I'll call it, a *perfect* relation.

Consider any mechanism used for a purpose. No matter how well contrived it is, or how well people use it, mishaps occur. Things always go wrong; and often, if possible, we build safeguards into our practices with the mechanism to prevent disasters. Here's a related point. Any mechanism for doing something usually must be learned; and, during the learning process, one makes mistakes. The better we get, the fewer the mistakes. Consider our ability to utter grammatical sentences. This, even if a capacity we are in some sense born with, must still be learned over time, and practiced. In fact, mistakes made during the learning process are significant indicators of the psychological processes a child goes through before acquiring a finished ability to utter grammatical sentences.

As an illustration of the point about learning, recall from Part I § 8 and Part III § 6, that our methods, devices, etc., for getting in touch with things always pose a danger of our confusing artefacts produced by the methods themselves with properties of the objects under study. In learning how to use such methods, therefore, one also learns to distinguish artefactual properties of the methods used from properties really possessed by what is being interacted with.

Prima facie, if reference for kind terms is supervenient on causal relations, or is constituted, in some sense, out of causal relations, then since such causal relations are exploited or forged by various mechanisms which can misfire, reference too should be something that can misfire.[17] But it seems our intuitions and practices do not allow referential mistakes of this sort. For we can attempt to get what we think about a stuff right (e.g. replace a wrong description with a right one); we can attempt to refine our causal relations with a stuff (e.g. get purer samples of it than the ones we have; get a stronger pair of eye-glasses); but we can't improve our references to anything (e.g. make sure every use of the term "dog" refers to a dog

---

17 Understand "mechanism" broadly enough to include our support system, e.g. nerves, eyeballs, etc., for our perceptual abilities.

Although the sentence to which this footnote is appended is generally true of kind terms, it's *not* generally true of proper names. My reference to Einstein is passed to me by other individuals, and it ends up referring to whatever the causal chain ends in (a bottle set asail in the past drifts finally into my purview). So what will be described shortly as the perfection of reference is no surprise with proper names. But where procedures are forged and refined to establish causal connections to instances of a kind, perfection of reference *is* a surprise.

rather than including the occasional unkempt cat). Most of what we do is open to improvement, most of what we do is imperfect. Our ability to refer, by contrast, admits of no improvement. A child, in time, learns more about apples, but in doing so she doesn't succeed in referring to additional apples – apples she didn't refer to before – by use of the term "apple." Children don't learn to refer better as they get older.[18]

Some may dispute this intuition: S picks up a genetically-altered pear, bites into it, and marvels aloud at the sweetness of "this apple." S *has* referred to the pear with the term "apple." Sure. But it isn't that the word "apple" as used by S now misfires and refers – at this moment – to a pear. A distinction must be made between what S, using the word, refers to, and what the word, in any case, refers to.[19] Perfection of reference is a matter of the reference of the word.

It might be thought that deviational failure in the mechanism of reference is definable in terms of the distinction just mentioned – a speaker's reference differs from semantic meaning precisely where referential mishaps occur. Maybe, but this doesn't by itself explain what needs explaining: a causal mechanism underwrites *semantic meaning* (that's the causal theorist's assumption), what the word *"apple"* refers to. So how come the latter mechanism (all by itself) doesn't misfire?

One might try to explain the perfection of reference by showing how it follows from the linguistic division of labor. This handles the irrelevance (to their capacity to refer) of the general ignorance of children and lay-persons about the referents of their terms; my confusions about elm and birch trees don't impact on the mechanism of reference because I and my actions don't contribute to the mechanism of reference for these words (although experts and their actions *do*).

This successfully explains the perfection of reference of proper names for an interesting reason. No matter *what* happens, reference succeeds, *if it gets off the ground at all*, because it's to *whatever* object the causal mechanism results in. (Malfunctions don't exist, nearly enough, because *if something* happens, it counts as a success.)[20]

This isn't going to work in the case of kind terms, because the *causal mechanisms* are too complicated: (1) in general, what's referred to is a class of items; and (2) the causal mechanisms used to reach these items are

---

18 One is reminded of what Descartes (1641: especially p. 175) said about the perfection of the liberty of choice; one is equally suspicious.

19 And the distinction *has* been made. See Kripke (1979).

20 Cases where reference doesn't get off the ground *at all* are cases where the mechanism yields two items or none. This could be seen as a kind of referential mishap – but I'm interested in a particular type of blunder that's ruled out: referring to *A* instead of *B*. It's why this kind of blunder can't happen that's been explained by the mechanism of reference for proper names being regarded as a success, no matter what it produces – provided it's one thing.

procedures, a heterogeneous bunch with only partially overlapping extensions. This makes it hard to see how referential success could be *whatever* the procedures (jointly) yield. Adjudication of some sort seems clearly needed where they disagree; and since the procedures are fallible and replaceable, this move forces reference to be *dispositional*.

I showed in § 2 why *that* gets us in trouble. Here's a reprise in the language in this section: just as my reference to gold is grounded in a potential causal relation between myself and experts,[21] so, too, cases of reference not underwritten by actual causal relations must be handled by arguing that "refers" operates via dispositions grounded in the criteria of terms. This works fine for criterion-immanent terms: their range includes items that fit the criterion, not items we *mistakenly* take to fit the criterion (and this explains the perfection of reference for such terms).

The strategy faces problems with criterion-transcendent terms, however. The dispositions needed in these cases have to go beyond ones grounded in terms of the criteria of words already in place, because these criteria can change: We need not just dispositions implicitly defined by criteria, but also dispositions *in us* for *changing* criteria. But § 2 yielded the claim that dispositions *like that* are *not* naturalistically available. If *every* term were criterion-immanent, there would be no problem. We would understand ourselves to be changing the term each time we changed its criteria.

As far as our intuitions are concerned, the suggestion that "refers" – at least with regard to criterion-transcendent terms – is underwritten by causal mechanisms, or by such mechanisms in conjunction with other resources, looks like a premature generalization from the case of proper names. There, the perfection of reference is grounded in our capacity to allow causal relations perfect freedom so that whatever (unique thing) they turn up is what is referred to. Reference thus is perfect: an all or nothing affair. But we cannot be so blasé about causal relations underwriting kinds.

Well, one can say, there's still the revisionist ploy left: forget our intuitions and the troubles they give naive naturalists. Such intuitions are psychological leftovers of a dark and superstitious time when magic ruled the word.[22] Instead let's mint a new language where the referential scope of our terms coincides exactly with our causal powers. I evaluate this suggestion next.

---

21 "Potential" because the linguistic division of labor involves deference. It's not that the items I refer to by the term "gold" are those *actually* recommended to me by experts. It's rather that, if necessary, I'll search out an expert and seek her advice with regard to a disputed item. In this case, the dispositional relation owes its existence to *actual* social relations – my belonging to a certain linguistic community.

22 It *does* seem, after all, that intuitions supporting the perfection of reference with regard to criterion-transcendent kinds are a significant part of what Putnam (1981: 3) has derisively called "the magical theory of reference."

§4

CAUSALITY AND REFERENCE
An analysis

I start with a review of some general properties of the collection of procedures (the methods of generating causal links between our terms and the things they refer to).

(1) They're an *open-ended* collection, for we're always developing *new* ones. Sometimes a "new" procedure is a fine-tuning of an old one, so it gives different answers in certain cases. But technological developments also involve inventing new mechanisms without obvious antecedents.

(2) Procedures, generally, are trivalent and vague. By "trivalent" I mean that there are cases where we apply them, and get the answer "Yes"; cases where we get the answer "No"; and cases where we can't get any answer at all. These last can arise one way, at least: we may be unable to apply the procedure in a particular situation (try biting into something located in the sun). By "vague," I mean that in certain cases a procedure may give no clear and unequivocal answer, or the answer may not be *stable* (repeating the procedure in the same situation results in a different answer).[1]

(3) Procedures *overlap* in their domains of applicability. For example, recognizing gold by means of one's senses gives the same answer on many items that more sophisticated approaches do. But it doesn't *always* agree with them.

With these facts clearly in mind, let's provisionally adopt the revisionist project: We drop our current language, and instead associate a unique term P_p with each procedure p, where P_p applies to any item p gives the answer "yes" to.[2]

---

1   Scientific practice, especially in engineering contexts, is often a matter of developing procedures, so the third set of answers is narrowed in scope and the first and second sets are expanded. See, e.g., Kateman and Buydens (1993).

2   It may seem that I'm blending together distinct sorts of causal interaction. We *test* for the

This is rather *like* classical operationalism (I'll call it "causal operationalism," hereon). Each term we have has a procedure associated with it. When we invent new procedures or modify old ones, we discard the old terms (if any) and introduce new ones.[3] Is this a bad idea? Yes; and for reasons close to, although not identical with, those which sank classical operationalism. The matter is delicate, however, and so the drawbacks of causal operationalism should be analyzed in some detail.[4]

What is the relation between scientific theory and procedures? Well, broadly speaking, we find the following: (1) current procedures are explained by scientific theories; and (2) scientific theories, both current ones and new ones, motivate the development of new procedures. Because the

presence of something, but we also *manipulate* things to get at other things, and some things we just *use*. Nevertheless, these interactions all involve procedures in my sense, for if an item fails to be what we think it is, it will fail to participate in the causal chain in the right way, a fact, presumably, that can be recognized. Tests are distinguished from use and manipulation, by being relatively easy ways to recognize things, as opposed to hard ways, and by being ways of recognizing things, which, when they give the wrong answers, do not lead to big complications (the *litmus paper* turns blue rather than *someone* turning blue). But this distinction, although important for applications, does not bear on the issues I'm concerned with.

3   It's worth stressing what just happened. In requiring terms with referential scopes identical to their causal scopes, I've found myself inadvertently reviving something rather like a long discredited philosophical doctrine. But this should be no surprise. Fatal to classical operationalism was its apparent reliance on an observation–theory distinction. But the doctrine was complicated, and had anyone at the time been interested in metaphysical issues about causality, rather than epistemological issues about verification, this might have been seen. For what's crucial about classical operationalism, from the contemporary vantage-point, is that operationalizing a concept really amounts to determining what the causal links between it and the world are; even more striking is that in suggesting the meaning of a term lies in the procedures operationalizing it, proponents of the doctrine were really presenting a kind of causal theory of reference. This is why similar issues (e.g. counterfactual conditionals) come up in the literature of both topics. See e.g. Fodor (1990b, 1990c); Hempel (1954).

By the way, sometimes the doctrine is called "operationism," as Hempel (1954) and Devitt (1991) do, and sometimes it's called "operationalism," as Boyd (1991) does. I adopt the latter and its cognates because they sound better.

4   One common objection raised against classical operationalism is that it attempts to eliminate theoretical entities, and such entities are required for scientific practice. See, e.g. Boyd (1991: 9). Also see Hempel (1954) for a very careful presentation of classical operationalism. On the nonexistence of a distinction between theoretical and observational terms, a distinction which philosophers felt classical operationalism required, see Achinstein (1965) and Maxwell (1962). Also see Putnam (1983c: 282–3). These sorts of objection fail against causal operationalism because the latter has no aspirations to eliminate theoretical entities, and no need for a distinction between theoretical and observational terms. (Recall the objection raised to Putnam's version of the permutation argument in Part III § 5: he failed to see that the causal theorist is concerned with operational constraints construed metaphysically rather than epistemically.)

causal operationalist has aligned *her* theoretical terminology to *current* procedures, she can easily handle (1), but not (2).

On (1): *First illustration.* The critic of causal operationalism might object this way: it's an important fact that our tests are fallible, that what they identify as something is not always that something (what we take to be gold on the basis of our senses, is not always gold). In indexing our scientific terms to procedures the way the causal operationalist wants us to, we've lost this important insight. Furthermore, a kind of explanation we like seems to have been lost. We might subsequently develop a certain procedure that gives "the right answer" in certain circumstances, which our previous ones didn't; that is, say, it recognizes that something isn't gold in circumstances where the previous procedures falsely identified the presence of gold. How are we supposed to say this if we don't have a term "gold" that picks out the natural kind?

Response. What talk of fallibility amounts to is only the observation that things which act alike in certain ways won't always act alike in other ways – that what fails to be distinguished by certain procedures *will be* distinguished by others. Construed this way, the insight available by means of talk of fallibility is still available. Call the golden stuff we recognize by means of our senses gold$_{sen}$, and call the golden stuff we recognize by means of a certain procedure t, gold$_t$.[5] Then the point about fallibility is simply this: not everything that is gold$_{sen}$ is gold$_t$, and for certain purposes gold$_t$ is better. Furthermore, there is no problem justifying the introduction of a new procedure u, and with it the term gold$_u$, say; for the items u recognizes as being alike are particularly useful for us: they act alike in circumstances where gold$_t$ just doesn't. Notice how talk of fallibility has dropped out of our explanation; this is because it's an idiom causal operationalists don't need for explanation. (This, anyway, is the claim I'm trying to establish.)

Second illustration. Suppose we silver-plate a spoon by an apparatus using a certain chemical solution the spoon is submerged in. Our ordinary explanation for why this process works is that the liquid has silver ions in it. Without access to natural-kind terms we are left with a method for coating spoons with silver$_{sen}$ by means of a solution (containing no silver$_{sen}$) without an explanation of why this works. In fact, there is no hope of an explanation – for by *definition* silver$_{sen}$ cannot be found in the liquid.

Response. The old explanation for why the spoon gets coated with silver is that silver (in another form) was there all along. The causal operationalist replaces this explanation with an explanation in terms of a law-like

---

5   For the causal operationalist, the appearance of "gold" in terms like "gold$_{sen}$" and "gold$_t$" is opaque. The two kinds named by these terms may be related in a lawlike way, but *not* because they are two kinds of *gold.*

connection between different substances indexed to different procedures.[6] The liquid in question may be something that can be recognized only by highly specialized procedures (so the liquid is not what we might call an observationally recognizable stuff). But where the proponent of natural kinds gives an explanation in terms of the law-like properties of one kind of thing, silver, with both observational and non-observational instances, the causal operationalist has an explanation in terms of the law-like properties of many kinds of things, some observational and some not.

This last illustration enables me to sketch out how traditional scientific theory can be replaced by the causal operationalist's new "procedurally honest" sort of theory without explanatory loss. Consider the following idealized, and historically false, example. Suppose our theory tells us that the weight of certain substances (carbon, nitrogen) will measure evenly in certain units because these substances are, as it were, "made up of" stuff (hydrogen) that comes exactly in such units, and never in fractions of them. This theory "determines" a certain set of measurement practices. But it isn't the only theory that does this. A theory that simply postulates law-like (periodic) connections between the weight of substances yields the same procedures.

Now here's a worry. It seems that the substitution by a theory with more "natural kinds" for another with less is not merely a matter of substituting the more complicated for the less complicated (as if that wasn't bad enough): in addition, a crucial explanation is lost, as the idealized and historically inaccurate example shows. For in one case, we have many laws connecting substances, without any explanation of why we have such laws. In the other case, these laws are explained in terms of underlying, and unifying, laws from which they can be deduced. If substances really are "made up of" hydrogen, then we can deduce the many laws in question from the single law about hydrogen.

The causal operationalist responds this way: it's misleading to claim we now have many laws connecting substances without any explanation of why we have such laws. For suppose we have two laws A and B, and we later deduce B from A. Although we now have an explanation of B in terms of A, we don't have any more of an explanation for A **and** B than we did before. So the sort of explanation this objection focuses on amounts to nothing except the increased deductive unity of our theories. No one denies the

6   But wait. Doesn't this fly in the face of Occam's razor? Don't we, all things being equal, prefer an explanation in terms of one substance to an explanation couched in terms of law-like connections between several substances? Sure. But why? I'll argue shortly that the only way to justify this reading of Occam's razor is in terms of considerations about the simplicity of our terminology. For the time being, though, I'll accept that shifting to causal operationalism costs us simplicity.

appeal of this; but it's misleading to suggest that explanatory power in some *other* sense has been lost.

Before turning to causal operationalism's flaws, I draw some interim conclusions. First, there is no loss of explanatory power – other than explanatory unification – in adopting the terminological reform causal operationalism calls for. Either we can still say the sort of thing we were saying before (albeit in a different way), or the sort of explanation lost seems to be an unneeded artefact of natural-kind terminology (e.g. talk of fallibility).

Furthermore, the new terminology urged on us by the causal operationalist has definite philosophical advantages over the old-style terminology. There are no metaphysical problems regarding what terms apply to, and how we manage to get them to so apply. This is indicated by the absence of deep puzzles about referential mistakes: how mistakes are made does not need an explanation. A procedure can certainly be applied in the wrong way – doing so amounts to applying a *different* procedure. *But* it's not possible for a procedure to give the wrong answer, for it's assumed in causal operationalism that a procedure applies to whatever it gives the answer "yes" to. That is, the procedure p itself (modally extended, if necessary) marks out the causal link needed for reference, and so the term P_p applies to any item the procedure p would give an affirmative result to.

There is another philosophical dividend to this story. The recent Occamic considerations raised against the causal operationalist can be turned to advantage by noticing a problem that has always faced the scientific realist: the need to explain how the application of Occam's razor to limit, where possible, the number of natural kinds can be justified metaphysically. There seems to be no connection between our methods for deciding which theories are true, when such methods make heavy use of Occam's razor, and what the universe itself might be like (why *should* the universe be simpler?). This problem does not arise for the causal operationalist since she doesn't shave.[7]

Now for the bad news. How does causal operationalism handle theory-generated changes in procedures, or the use of theory to develop new procedures? I can mute the problem forthcoming a little by reminding the reader of the weaknesses in theoretical deductivism, something the challenge draws a lot of its impact from. Recall[8] that the refinements of procedure that take place in the laboratory are refinements that frequently take place, and frequently must take place, *independently* of theoretical considerations, simply because tractability problems prevent us from applying theory in a direct enough fashion to yield predictions about the experimental situation. The simple desire to make a procedure operate in a uniform way is often enough by itself to fuel a successful modification of the procedure.

---

7 The problem is that Occam's razor seems to be an *a priori* constraint (read: stipulation) on things we have no right stipulating facts about. (This problem has come up before: § 2.)

8 Part I §§ 3 and 6.

But I'll concede that, even if theory-generated changes among procedures take place more rarely than commonly thought, still, that they take place at all seems enough to produce an objection to causal operationalism: We use theory, and the terms that appear in theory, to motivate changes in our procedures. Can the sorts of theories available to the causal operationalist do this? No, and for a fairly obvious reason. Our standard language allows us to refer to things apart from any procedures we may have for picking out these things, and this means we can formulate regularities that apply to things independently of what procedures are available. Consequently we have terms we take to apply to new situations where old procedures for these terms may not apply. But if all the kind terms in our language are indexed to procedures already existing, then when faced with a situation where old procedures don't apply, we have no terms that describe anything there, and no new regularities to guide us in generating new procedures.

So, imagine we have a procedure p for recognizing the presence of a chair in daylight (we look for, say, a certain shape). Now imagine we're in a room without daylight. Even with flashlight in hand, we have no idea what to do. For by definition there are no chairs$_p$ in the room. And so nothing we know about chairs$_p$ will be of use to us. There *are* chairs$_f$ (chairs-that-may-be-seen-by-flashlight); but, if we discover this fact, it can be only by sheer accident.

Of course, a great deal of what we discover *is* by sheer accident. And sometimes the theories that guide us in our exploration of the world around us are ultimately heuristic, in that the discoveries made by such means can discredit the theories themselves. So the causal operationalist can argue that the benefits lost are purely instrumental ones. Theory (by means of the language it's couched in) fires the imagination by going beyond what we're causally licensed to talk about, and this has a payoff. But even if this is conceded (for the causal operationalist will stress that the failing here only involves the development of new procedures or the finetuning of old ones; theories couched in casually acceptable idioms can always explain *current* procedures, as we've seen), there is still a burden on the causal operationalist to supply an equally useful psychological mechanism for designing new procedures the way that current theories help us to. I've no idea what this could look like.

There's yet another big problem: the language the causal operationalist urges on us is *terminologically impractical*. Motivated solely by the desire to resolve the question of how we interact causally with what we refer to (this is why reference is constituted entirely out of procedures), not a thought was given to what a mess such single-mindedness would make of our *talk*. Consider the practical constraints on talk. We need our terminology to be long-lasting and small (relatively speaking) for it's work to master new vocabulary on a regular basis, and it's irritating to constantly discard vocabulary. But look what causal operationalism saddles us with: each time a procedure is modified (and procedures are modified constantly), the term indexed to that procedure goes; each time a new procedure is invented (and new procedures are invented constantly), a new term is called for.

Furthermore, Putnam's linguistic division of labor manifests itself in a particularly ugly way here, since particular procedures are often the domain of specialists. As a result, the vocabulary of the scientist will be largely divorced from that of the lay-person, and the vocabulary of each individual scientist (even in the same field) will be distinctive. Nevertheless, all this massive vocabulary will be connected in law-like ways that pose monstrous bookkeeping tasks, because items grouped according to certain procedures will have law-like relationships with items grouped according to other procedures. Think of all the procedures we use to recognize and deal with gold (according to our current story) – these alone will degenerate into a giant spiderweb of interconnected terminology that only a well-paid bureaucrat could love.[9]

Causal operationalism is therefore unacceptable. Not because it doesn't do justice to our current scientific practices, but because (1) it stifles our capacity to use scientific theory to innovate procedures, and (2) the language it requires is intractable.

Occam's razor is an appealing objection to make against causal operationalism, as I briefly noted earlier. One can argue that, all things being equal, it is better to have the simplest scientific theories and explanations we can. So theories that use fewer natural-kind terms, and consequently have laws which are deductively more unified, are better theories. The causal operationalist has a response, of course (apart from her worry that the razor raises philosophical puzzles about how it's to be metaphysically justified): "All things being equal," includes our capacity to tell the appropriate (naturalized) story of how we refer to instances of the natural kinds postulated. And merely focusing on simplicity as a justification for the razor doesn't guarantee this.

But, notice: *pace* the motivations for causal operationalism, the scientist never concerns herself with this sort of worry: scientific methodology is never concerned with guaranteeing that the referential relations of newly minted scientific terms are naturalizable. Rather, the reference of a term is always taken for granted, except for the caveat that the objects involved may not exist at all. This needs explaining.

9   There is yet another problem, although it's not quite as important as the above. Causal operationalism complicates our language by requiring a three-valued logic. This makes things *only* a bit worse off: the logic would be strong Kleene – a fairly well-understood logic with some nice properties.

§5

TRANSCENDING PROCEDURES

Terminological practicalities refute causal operationalism, and with it, any hope of explaining how we refer in ways compatible with naive naturalism. I turn now to describing how our referential practices actually work, and seeing how I can make philosophical sense of it.

Here are some preliminary methodological points. First, recall that I've elicited a number of intuitions about reference having to do with its elusivity, its perfection, with the distinction between criterion-immanent and criterion-transcendent terms, etc. Part of the evidence for the picture I like of how reference works is that, once it is in place, it will show where these intuitions are coming from.

If we could have stayed with the sort of language required by causal operationalism, then there would be no philosophical problems about how we refer; that is, there would have been no problems about how to naturalize reference, no problems with justifying Occam's razor, etc. It's the solution to the problem of how to generate a nomenclature psychologically supportive of our theorizing practices, and not too complex to use, a solution implicitly at work in our current referential practices, that gives rise to the sorts of philosophical problem about reference already seen. But this point leads to another (philosophical) solution to these problems, as I'll show.

Finally, I use the language of causal operationalism as a kind of creation-myth. I imagine, that is, starting (in a simple enough linguistic context) with referential customs compatible with the strictures of causal operationalism, and then seeing how far we must be driven from these customs by the difficulties raised in § 4. Equivalently, I'm using the position of the causal operationalist as a foil for various tempting revisionist projects; the language of causal operationalism is an upper limit on what's acceptable to the naive naturalist, but is below the lower limit on what is terminologically possible.

So, how are the terminological problems that causal operationalism faces solved? Answer: combine *groups* of procedures under the rubric of single terms; attempt to hang onto those terms despite changes in the procedural personnel associated with those terms; and do so in a way that maximizes both the deductive unity of the empirical sciences and the set of simple generalizations we can articulate. This way we solve both of our practical difficulties with terminology: we minimize the vocabulary as much as we can (Occam's razor), and we minimize also change in that vocabulary.

How far this move can be implemented is, naturally enough, an empirical matter. Some suggestions ("let's call everything 'water'") don't pan out. Some of our vocabulary ("witchcraft," "caloric fluid") just has to go. But we've been relatively lucky, all things considered. Not only have we been able to tie together procedures that differ from each other only minimally (procedures, that is, which are improvements on each other), but it's been possible to make even deeper associations between procedures than anyone could ever have imagined. This is reflected, for example, in the fact that all chemicals are compounds of a finite number of elements.

I suggested that the move to make is to group procedures under single terms. But this leaves a lot of options. In particular, it leaves open a choice about how to treat the extensions of such terms when the procedural personnel changes: Such changes can be either strongly retroactive, or, at most, weakly retroactive. Which is suitable? The answer, naturally, is that different terms should be treated differently: Some should be criterion-transcendent, and others criterion-immanent. But on what basis?

Consider the collection of terms operating in the natural sciences. One aim we certainly have in modifying the procedures for these terms, as mentioned, is greater deductive unity in the theory and application of the sciences, for greater deductive unity yields greater power for the theory. To this end, it's desirable to sharpen law-like connections between procedures both within the umbrella of particular terms and across such terms. But we can't do this unless we're allowed to rewrite the extensions of natural-kind terms. If certain procedures previously associated with the term "gold" pick out a certain extension, and we subsequently change the procedures, we can't be beholden to the previous extension. It must be repudiated, for otherwise the term "gold" will not be suitable for appearing in general and (relatively) simple generalizations; any such generalization will have to cover not just the instances we have refined out of the previous extension of "gold" – but all the old instances as well, even if *all* such items are not easily covered by a generalization.

This argument motivates only weak retroactivity, not the strong retroactivity that natural-kind terms, among others, seem to have. For the considerations raised show only that we don't want "gold," and other terms that operate the same way, to *currently* apply to what we previously took them to apply to. But this doesn't require that we take "gold" to *never have* applied to what we previously took it to apply to.

What other considerations are pertinent? This, at least: we also want the laws that our natural-kind terms are couched in to apply (at least programmatically) to everything we can get them to apply to. In particular, this means, especially in the context of physics, that they should apply to the very procedures that have been discarded: We want to explain what such procedures picked out, and how they picked it out (through what mechanism); and we want such explanations couched in our *current terms*. This is enough to force strong retroactivity; we may find that to explain how an earlier procedure associated with "gold" operated, we have to notice, say, that it tended to be unduly sensitive to minute amounts of copper, and so rejected certain items (having a rather high percentage of gold) as containing gold because of this. This approach to explanation would be vitiated were we to treat the earlier procedure as having picked out gold exactly when and where practitioners using it thought it did.

Contrast the explanatory requirement I've placed on criterion-transcendent terms with the requirements on other terms. Legal terms, for example, are often weakly retroactive because there are simplicity requirements on the use of such terms quite analogous to the simplicity requirements on natural-kind terms. But such requirements in the legal case are usually restricted only to current usages, not previous ones, and this is why intuitions about such terms rarely require anything more than weak retroactivity.[1] Thus, we may introduce simplifications in the procedures for notarizing documents; but such procedures, and the terms they're associated with, have no bearing on previous procedures associated with the terms, and are not responsible for them, explanatorily speaking, in any way.

Now consider a somewhat different point. The general move encapsulated in our referential practices is to shift all the messy issues having to do with procedures out of the topic of reference (out of metaphysics, that is) altogether, and into epistemology; the result is to leave as the theory of reference something clean and entirely suitable to incorporation in a (Tarski-style) theory of truth.

I illustrate how to do this by shifting the perspective on criterion-transcendent terms. These terms, recall, are ones where changes in procedures are strongly retroactive. But describe them in this equivalent way: criterion-transcendent terms are ones where the connection between what they refer to and our procedures for picking out what they refer to has been loosened. Officially, procedures are not treated as constitutive of reference, for the references of such terms *transcend* any tests or procedures associated with them.

---

1   "Murder" is an interesting counter-example. Intuitions that lead to classifying it, and a number of other moral terms, as criterion-transcendent can be explained in a way analogous to how we explain intuitions about the criterion-transcendence of natural-kind terms: Moral idioms participate in *explanations of* shifts in the procedures and criteria associated with, among others, legal terms. E.g. the explanation of why a certain law was eventually rejected might turn on its immoral consequences. I cannot go further into this very interesting topic now. Notice, though, that this sort of explanatory application of moral terms prevents their being isolated in the sense of Part I § 7.

There are two sides to transcendence. First, the procedures affiliated with a term are fallible in particular cases, and in fact eliminable wholesale: we may find over time a term failing to retain even one procedure it started out being associated with. Our procedures, in a phrase, are epistemically *slighted*. But this epistemic slighting must be understood in a manner that places obligations on us epistemically. Although we need not, and may not, be able to explain ahead of time exactly how or why such procedures fail to pick out what is referred to, an "explanation" for this *is* required (in the fullness of time, anyway).[2] Of course this doesn't mean we must guarantee, ahead of time, that such explanations will be forthcoming; that is impossible.

But often such explanations *are* available. In the fullness of time, we explain the failure of our hands-on methods for recognizing gold when it comes to sea water by the fact that the gold there is too minute to detect by these methods. We may explain also the failure of prototypes of certain devices to pick up what their descendents notice with ease by invoking human error, malfunctions, sheer crudity, etc.

Notice that what would have been seen by the causal operationalist as a metaphysical issue (the referential relations of our terms to the world are changed) has been re-described as an epistemological one (certain sorts of "mistake" have been made).

The other side of transcendence is the metaphysical purity of "refers." I mean by this several things:

(a) Reference is *elusive*: it can't be constrained by, supplemented with, or defined in terms of our actual practices, however idealized or refined (with regard, say, to ultimately completed science).
(b) It's a perfect relation (§ 3).
(c) The references of our predicates are taken to be bivalent and unchanging, despite nothing either bivalent or unchanging underlying them.[3]

It's worth pointing out, by the way, that these properties are ones we have seen to arise, intuitively speaking, with respect to reference, and all of them,

---

2   Where's this requirement coming from? I.e. if shifting on our procedures is motivated by a desire for neat generalizations, where does "failure" come in? This way: we want to explain, given our new procedures, and what's picked out by our terms in respect of them, what our old procedures were picking out. In doing so, we reclassify what the old procedures pick out and in terms of this reclassification, both describe them as mistaken, and explain why they yielded the results they yielded. Along the way, we may also see how, in terms of our new reclassification, the old procedures prevented the exemplification of one or another generalization we now accept.

3   Vagueness, *semantically speaking*, vanishes, for it arises from both the vagueness of particular procedures themselves and from the clash of different procedures for the same term. But all this is reclassified as epistemic. See § 6 for further discussion of this.

it turns out, are due to the metaphysical purity of reference, that is, our reclassifying of issues having to do with procedures out of the topic of reference and over to epistemology.

Here's an illustration. The Greek term for gold is coreferential with the Spanish term for gold (during, say, the Spanish occupation of South America) and with what we refer to by "gold" today. (We say things like: "The Greeks did not know there was gold in sea water," rather than "The Greek term for gold did not refer to the gold located in sea water.") We don't treat the term "gold" (modulo translation) as an indeterminate one, the reference of which is progressively fixed as procedures are sharpened and developed. Instead of reference being a changing many-splendored thing, it's an unchanging and perfect relation; and it's our knowledge of things that is changing and many-splendored.

Let's develop this theme further. Field (1972) raises an objection to Tarski's approach:[4] Tarski's approach to truth is seriously incomplete, since the notion of reference is given only by means of a list (for a first-order language without function symbols, and using a finite vocabulary, say) of the references of the individual constants, and the satisfaction-conditions for the predicates. Field (1972, pp. 362–3) says:

> Now, it would have been easy for a chemist, late in the last century, to have given a "valence definition" of the following form:
>
> (3) $(\forall E)(\forall n)(E$ has valence $n \equiv E$ is potassium and n is $+1$, or ... or E is sulphur and n is $-2)$
>
> where in the blanks go a list of similar clauses, one for each element. But, though this is an extensionally correct definition of valence, it would not have been an acceptable reduction; and had it turned out that nothing else was possible – had all efforts to explain valence in terms of the structural properties of atoms proved futile – scientists would have eventually had to decide either (a) to give up valence theory, or else (b) to replace the hypothesis of physicalism by another hypothesis (chemicalism?). It is part of scientific methodology to resist doing (b); and I also think it is part of scientific methodology to resist doing (a) as long as the notion of valence is serving the purposes for which it was designed (i.e. as long as it is proving useful in helping us characterize chemical compounds in terms of their valences). But the methodology is not to resist (a)

---

4   The objection is one that applies to any formal approach to truth, e.g. Herzberger's, Kripke's, and so on.

and (b) by giving lists like (3); the methodology is to look for a real reduction. This is a methodology that has proved extremely fruitful in science, and I think we'd be crazy to give it up in linguistics.[5]

My view need not concern itself with this issue. Physicalism, as Field understands it, calls for a characterization of reference in terms of the actual procedures – physically construed – that (currently) attach us to the objects we refer to (to the extent that they do so). But these messy procedural facts are not semantically relevant. Reference is *elusive*, and Tarski's notion of reference (as with truth) cannot be constrained by or supplemented with causal conditions, or replaced altogether by a notion characterized in terms of the procedures. This *doesn't* mean that we can't explain what's going on in reference in a way compatible with naturalism; we can, just as we can describe our current methods for gathering truths without being able to incorporate a description of those methods into a characterization of truth. We can describe the procedures (at a time) that enable us to contact objects, but cannot include a description of these procedures in a characterization of *reference*.

The metaphysical purity of "refers" and the epistemic slighting of our procedures work together so we can embed our terms in simple laws and generalizations. Now we can escape the bizarre results procedures often saddle us with by denying that they bear on what our terms *actually* refer to, and relegating the issue of the source of the "deviation" to epistemology. From the point of view of the procedures – from the point of view of what we *do* – this is invoking a referential relation out of thin air, but this is not recognized as such in what we *say*, that is, in the implicit semantics of our terms.

The semantic naturalist might be worried. Why doesn't the move contemplated here simply violate naturalism outright; why isn't it the postulating of a metaphysical relation (between us and what we refer to) which can't be explained in terms of anything that we do, causally or otherwise? Why isn't the acceptance of such a relation therefore the acceptance of a highly suspicious occult notion?

The answer is this: in one sense, there *is* no relation between us and what we refer to. Or, rather, the only relations are the ever-changing and flexible causal ones we construct on a regular basis, but which we treat as (purely) epistemic onslaughts on what we successfully refer to already.

---

5   One can wonder whether Field has characterized scientific practice correctly, especially when it comes to a special science like semantics. Apart from his strong reductionist viewpoint, the kind of view I've opposed in Part I § 5, McGee (1991: 82–6) argues that there are many sciences in which appropriate methodology *does* involve lists much like the above, and he singles out lepidoptery as an example.

Stalnaker (1987: Chapter 2) notes that a Field-like point may be generalized to the logical connectives used in a Tarskian definition of truth as well. Here, too, we have a list that, if Field is right about scientific methodology, must be replaced by a genuine reduction.

But the nonexistence of such a relation doesn't stop us from talking *as if* such a relation exists. And why should it? Such talk, though, involves a certain kind of stipulation. We treat previous usages of certain terms as coreferential with current usages of these terms, even though the causal relations underlying such terms may have changed greatly in the intervening period. We handle the differences by means of the epistemic idiom, and say we were wrong about certain (many) items we previously took the term to refer to. There is no harm in this kind of stipulation, and a lot of convenience; which is why we do it. So our talk of reference doesn't place a heavy burden on the semantic naturalist, for she doesn't have to explain what "stuff" the referential relation is made up of. Rather, she only has to explain why we find it convenient to talk the way we do, and to expose exactly where the causal relations between us and the objects we refer to arise.

This provides an explanation for something noticed earlier. Recall (end of § 4) that scientific practice is not concerned with whether coined terminology is naturalizable (although there is concern with whether it succeeds in referring at all). Now it's clear why: the scientist wants smooth generalizations that can be applied easily. As long as a particular term remains in his lexicon, he can shift the procedures and the criteria for the term in ways that yield such generalizations, while continuing conveniently to speak as if he referred to the same thing all along. Resources for reference, therefore, are not an issue because they're treated as an epistemic matter. On the other hand, the successful production of nice generalizations may call for such radical changes in procedures and criteria that it doesn't make much sense to treat subsequent empirical terms as referring to the same things previous terms referred to; and in this case we introduce new terms and drop old ones.[6]

This helps us understand what might otherwise seem to be a weird fact: when we coin natural-kind terms, it looks like we almost always succeed in picking out natural kinds for them to refer to. This might seem to give Lewis' approach an edge over his opposition (*principle*: methodological miracles invariably involve presuppositional staging).[7] But there are several factors creating this impression that I now have the tools to explain.

First, we tend to forget we *don't* always get it right. Sometimes terms are

---

6   A quick comparison with the case of pure mathematics: because of our tendency in pure mathematics to keep as many mathematical results as possible, it's rare to find mathematical posits that are denied existence (as opposed to being neglected). On the other hand, generalizations in the empirical sciences are kept, too – it's just their status that changes: they may become approximations of other true generalizations (e.g. Newton's laws) or context-dependent results (e.g. that objects in the presence of frictional forces eventually stop moving). The difference is that empirical terms are linked to the world by procedures. And if we change the procedures for a term too much, or discard them altogether, we are loathe to retain the term we have so mutilated.

7   Recall the discussion of elite classes in Part III § 5.

just quietly dropped. That is, the *complex* of theory and procedures is dumped – and the procedures (or their descendents) worth retaining are embedded in our vocabulary in some other way entirely.

More significant, though, is the point raised several times earlier, that we often *retain* terms that *don't* pick out a "natural kind" – at least not in the pertinent scientific sense. This is especially the case with general terms applying to animals. We (i.e. non-specialists) use, for the most part, a shallow (although still fallible) list of phenotypes to identify the application of a term. That is to say, we rely on certain properties that the "kind" of animal exhibits to determine whether a term applies to an item or not, rather than its actual genetic history, or detailed phenotypic properties (that we don't have easy access to). Biologists use such terms, since they're common usage, although they necessarily introduce technical terms to mark out different genotypes as well. But many terms that pick out nothing genetically natural are in no danger of being dropped from our language, terms such as "rabbit," "porcupine," "ant," and so on.[8] On a view that makes "natural kind" – in the scientifically respectable sense – the backbone of a causal theory of kind terms, such terms are puzzling. On a view like mine, they're not: for the explanation of why natural-kind terms (and criterion-transcendent terms, generally) have no necessary and sufficient conditions is *not* that they're semantically structured to pick out metaphysical *kinds*, but rather because the conditions they do have are epistemically slighted. And they can have this property regardless of whether such a term (even with modifications in the procedures) will find a place in (one or another) science.[9]

In fact, the notion of "natural kind," in so far as it stretches beyond the scope of our procedures (at any one time), is as much a product of *talk* as "refers" is (since it's the extension of that relation). It makes our lives easier by smoothing our vocabulary, but does no actual work in picking out what our terms refer to.

Our tendency to diachronically identify terms that have rather different procedures and criteria, although stipulative in one sense, is not an arbitrary matter: identified terms have a family resemblance in their procedures and criteria. Such a resemblance is elicited not in a hard and fast manner, but in a soft way that can lead to controversy as to whether, indeed, the terms should be referentially identified. Ultimately, these sorts of controversy are

---

8   For those who like biology, but cannot take it raw, nice examples of evolutionary convergence – the emergence of species with similar traits, although dissimilar genetic histories, may be found in Dawkins (1987: Chapter 10).

9   But why don't such terms shift in status from criterion-transcendent to criterion-immanent when it becomes clear that they are no longer operating in pristine science? The reason is that they still operate explanatorily with respect to gross regularities (e.g. "Grabbing a porcupine without gloves will hurt"), and gross regularities, as I showed in Part I, cannot be replaced wholesale by scientific law.

grist for proponents of incommensurability. I see the issue this way: there are definite facts about procedures associated with terms, and the theory surrounding such terms; but these facts do not necessarily compel any particular stipulative identification between an earlier term and a later one. Stipulative identifications are not arbitrary, however, because they must yield the best smooth overall set of generalizations.

Here are a few illustrations of when we find it natural to treat two terms as coreferential even when the procedures associated with them, and the theories they appear in, differ. We often use hindsight to justify what we take earlier usages of a term to actually refer to. If a term was purely theoretical at first ("gene," for example), and even if the theory was vague, we take the term as referring to what our current term refers to.[10] If there were procedures associated with earlier usage which we have subsequently discarded, but (vague) theory was more or less in line with ours, we take the term as referring to what we currently refer to with the term. If the theory was all wrong, but many of the procedures were similar to, or were less sophisticated versions of, ones currently in use, we often take the term as referring to what we refer to currently. Finally, if a term A was originally used by us in a restricted context to successfully pick out (more or less) something we currently use A to pick out, but *neither* the procedures nor the theory in use at the time do the job in our (current) broader context, we take the term A as one which previously referred to what we currently take it to pick out.

One last observation about this matter. The fact that we tend to stipulatively treat terms as coreferential, even when the procedures and criteria for such terms have changed, has an important corollary. This is that we often stipulatively treat terms as coreferential even when their ostensible extensions are rather different. That is, *pace* straightforward causal theories of reference, we may treat two natural-kind terms as coreferential even when they do not have samples in common.[11]

---

10 On the other hand, if it turns out that too many *different* things live among what the original term was taken to apply to, the term may be dropped for more specialized terminology. This seems to have happened with "gene."

11 I am skimming the surface of a rather deep issue. These decisions of when and when not to identify terms as coreferential are not a heterogeneous collection of *ad hoc* intuitions. Rather they are motivated by the principle: make the referential apparatus as independent as possible of the ignorance or knowledge of its users. I can't go further into this matter now.

But I *can* add that another indicator of the transcendence of reference is that our terms *and* the truth-theory containing them really *are* taken to be *trans*theoretical: that is, in many circumstances we take a term to be referring to the same items even though the theory about it has changed (perhaps even drastically). In saying this, I don't want to give the impression that everyone is naively taken in by our collective insistence that certain terms corefer. As I have already mentioned, it's when interpreting different conceptual schemes that the transcendence of reference creates the most strain. Historians of science,

I want to conclude this section by pointing out again that I am certainly not the first to notice (some of) the properties of the referential relation that I've labeled its "transcendence." I've previously discussed Putnam, and Rorty on this matter. But let me make an observation now about Blackburn (1988). He notices what he calls the "normativity" of reference in the case of proper names, but draws a conclusion bleaker than mine. Since no definition of "reference" is forthcoming – no list of necessary and sufficient conditions for referring – he assumes that it follows that there is no "nature" or "essence" of reference for the theorist to uncover. Furthermore, he draws the conclusion that the absence of such necessary and sufficient conditions renders impotent "any attempt to found a realism *upon* a theory of reference."[12] He takes this last claim to follow from the fact that the absence of such conditions leaves no way of separating our methods for interpreting what speakers mean (what they are referring to) from our methods of determining the truth-values of their sentences.

This last worry, with its implicit invocation of the (Davidsonian) principle of charity, is a topic that must be left for another time. But let me address now the worry about *realism*. As I have stated, we can be fooled by the transcendence of "refers" into thinking that reference "really is" something over and above our procedures for picking out referents. In this case, "refers" truly looks "elusive": our procedures *are* indicators of what our terms refer to, but there seems to be more at work. Furthermore, it is that "more" which seems to do the crucial work: it eliminates all the funny cases and mistakes that slavish attention to the answers our procedures actually give would drag in. So not only is there more in reference than meets the operationalized eye, but this "more" is the crux.

But this is all (procedurally speaking) an illusion. Talk of reference over and above what is licensed by our procedures is just a semantic ghost which makes our verbal life easier by cleaning up our *talk*. Where *it* (talk of reference) seems to be firming semantic matter, enabling us to discard the awkward results of procedures, in fact the labor is being done by the lowly epistemic idioms, our desire to bring all the procedures associated with a term into congruence, and our desire for clean straightforward generalizations and a clean bivalent logic. In fact, how epistemically respectable our talk of mistakes and mishaps (in light of what our terms "really" refer to) ultimately reveals itself to be turns directly on how well (over time) we pay off the methodological debts incurred by this talk of errors and malfunctions. We should (over time) be able to explain, or at least correct, enough of them

in particular, are sensitive to how much theory-change and procedure-change alter our grip on the world, despite the soothing illusions engendered by our assumptions of coreferentiality. Hence the appeal of talk of incommensurability for philosophers who have studied the history of science.

12 Blackburn (1988: 192).

to make continuing the project (the theories with their particular vocabulary and laws) worth it; and, again, this is because generalizations and classifications are supposed to apply retroactively.

In short, naturalized realists and irrealists must compromise. The naturalized realist hopes for a naturalistic underpinning of our referential practices. He wants to see how, whatever our terms are taken to refer to, a causal mechanism shows how our use of the term connects to the items referred to. This is impossible. What he must settle for is a reference relation that is (partially and tentatively) underwritten by fragmentary but genuine causal relations to objects actually in the world.

The idealist, on the other hand, has a picture where talk of reference is, indeed, a projection onto the world. She uses arguments that put all our terms on a par: the world, as it is in itself, is indescribable (except from within our own conceptual scheme, with its made-up projections), and the world exerts its force on language so that there is no way to separate out where we're making it up and where we really are causally interacting with the world. Indeed, talk of causality, talk of reference, is taken by her to be something going on *within* our theory which has no metaphysical purchase on the world, as arguments involving proxy functions and alternative mappings of our terms onto the world indicate. During some moods, she even goes so far as to suggest that there is no world at all, or that such world as is left over is well worth leaving behind (or some witty variant thereof).

Instead, the idealist gets the fact that the reference relation (as a whole) cannot be reduced to something naturalistically acceptable. She also gets the concession that the extensions of our terms are (partially) "made up." But she must accept that our terms are *not* all on a par. No simple global arguments for linguistic idealism are available; instead, we see science laboriously extending what our terms are connected to, and we have the capacity at any time (if we are willing to study the pertinent sciences) to literally measure the purchase a particular term has on the world, by marking out the causal sweep of the procedures associated with that term.

In any case, both the realist and the idealist should realize that we're going to go on talking as we always have, and the point is to understand why we do it.

§6

TRANSCENDENCE AND ITS
DISCONTENTS

I've argued for a kind of split between what we do and what we say – and I've claimed this split explains much of the philosophical perplexity surrounding reference. My discussion of the issues, however, may have given the impression that when we step back and survey the situation we can escape the split: that we can, in a neutral fashion, describe our procedures, describe our referential talk, and see clearly how they don't match up. But if philosophy and its vocabulary is continuous with the vocabulary of science and ordinary life, this split should not be so easily escaped.

It isn't. Although I've spoken of procedures, my discussion of them has been informal enough to mask the language procedures must be described in – the full-blooded one I've attempted to motivate without begging the question against those with naturalistic scruples. To describe procedures is to describe what we do, or what devices (designed in such and such a way) do, and the language used is our ordinary one with its mixture of commonsense terminology and scientific nomenclature.

But if the argument in § 4 is right, there's no other option. Since the language contemplated by the causal operationalist really is intractable, we can't adopt it (even during a temporary attempt to be even-handed). Despite the context the argument must be mounted from (that is, from the language our current conceptual scheme is couched in), I don't think it, even in a twisted way, begs the question against the causal operationalist. Rather, it's a proof by exhaustion, the exhaustion of that philosopher who actually adopts the sequence of languages urged on us by causal operationalists.

Nevertheless, we can't lose sight of the fact that the language we couch our insights about procedures in is one whose vocabulary simultaneously sprints past those same procedures. The split goes all the way up the scientific hierarchy, and invades the ivory-tower enclave philosophy hides in.

A way to see this is to notice how certain traditional philosophical problems arise because of this split. It may seem that philosophical puzzles generated by the topic of reference are all variants on how badly the program of naive naturalism fits in with our referential practices. But that's not correct. Other puzzles arise from the seams in our referential practices –

points at which the way we have structured the way we talk about reference, mistakes, and so on, and the way we go about interacting with things in the world show strains. And it is these strains that philosophers have been sensitive to.

Let me give some illustrations of what I have in mind. First, consider the topic of mistakes. Mistakes come in two grades: robust and wan. Examples of robust mistakes are the misapplications of well-defined procedures. Should someone read-off information from a dial in the wrong way, or add the wrong chemical to a solution, and in this way yield a false positive, the mistake is a robust one. On the other hand, the claim, let's say, on the part of an ancient Greek that there is no gold in the sea, is a wan mistake; for the mistake is made in relation to procedures for which no possible corrections are available. Wan mistakes are the stuff of scepticism, the attempt to take the possibility of mishap as far as the imagination will allow ("perhaps we're all brains in a vat," "perhaps we're all being fooled by an evil demon"). And it's precisely against taking wan mistakes seriously that the causal theorist rails when he or she argues that the brain in a vat is not wrong but *right* to claim it's not a brain in a vat.[1]

Let's develop this example a little. We motivate the student in an Introduction to Philosophy course to accept sceptical scenarios by pointing out how often we make errors (and modify our views about what is going on in the light of them); she is taught how much procedures or criteria can change with criterion-transcendent terms.[2]

Also, notice how like ordinary shifts in our criterion-transcendent terms certain sceptical scenarios are: the brain in a vat has never been outside its vat of nutrient fluids (it was grown there, let's say, from an initial group of cells). Imagine also that the computer program generating the "virtual world" this brain lives in happens, by accident, to create a world that looks identical to ours. The odds of this happening, of course, are quite low – but it could happen – and the assumption makes sure the use of the terms utilized by the brain in a vat are not causally connected to referents outside the computer-programmed world. Now suppose that at a certain point the brain is removed from the vat and put in a human body which, again by accident, looks exactly like the one it thought it had before.[3]

The former brain in a vat would have the following intuitions (imagine this situation is described in a piece of science fiction, and notice that what follows is exactly how things would be described): before, while still a brain

---

1 Putnam (1981: Chapter 1).
2 Very little is needed to convince students that sceptical scenarios are possible. I speculate this is because the average 18-year-old has spent so much time being corrected (rightly or wrongly) by the adults around her.
3 This last phrase is tendentious, but I don't see how to describe the situation non-tendentiously, except at great length.

in a vat, it had thought it was interacting with chairs, trees, people, and so on, but now it knows that all of these things were just computer-generated illusions. In saying these things were not real, which is something it undoubtedly *would* say, it would not be saying they didn't exist at all, for indeed, computer images *exist as much as anything else*. Rather, it would be saying that all the things it *thought* were tables, chairs, people, and so on, *weren't*.[4]

Part of the intuition here is that the former brain in a vat is reasonable to keep its previous terminology. Indeed, the practices of ordinary science and ordinary life, where it's commonplace to shift (perhaps drastically) on the samples, procedures, and criteria associated with natural-kind terms, and, indeed, to acquiesce in the stipulative coreferentiality of terms used in a previous context with terms in a new context, explain *why* intuitions like these (so useful to the sceptic) arise.

On the other hand, we can make the student aware of how dependent our terms seem to be on the context in which they are learned; that is, that what they mean and how we understand them – our abilities to use them and indicate things around us by means of them – are due entirely to how we were taught to use them.[5] For this purpose, somewhat different scenarios are invoked. For example, one imagines, as Putnam does, a case where there only are brains in a vat, the vat with its nutrient fluid, and the computer (the whole thing, let's say, having come into existence spontaneously). In this case, one elicits an intuition that the terms used by the brains actually refer to what they seem to the brains to refer to, so that when a brain cries out proudly, "I am not a brain in a vat," it's *right*. Intuitions go this way here because in setting up the example, we've implicitly ruled out the possibility of wan mistakes (no getting out of the vat). Tweaking sceptical scenarios in

---

4   I should add that this scenario has the best shot at giving the traditional sceptic what he wants: a scenario where our intuitions support the idea that the subject of the scenario is wrong about nearly everything. For although the subject of the scenario can maneuver about in its new world, and although homonyms of nearly every generalization it believed still hold (e.g. "No chair is an animal," "Humans need food to survive," and so on), it could be argued that it didn't *know* any of these generalizations because it wasn't justified in its beliefs in any of them (justification turns on being right about particular instances of these things, and it wasn't right about any of those). Notice, though, that a simple nod to the sceptical scenario itself won't do the job for the sceptic here.

5   This does not require anything quite like causal considerations, by the way, as the anticipation of Putnam's argument on the part of Bouwsma (1949) makes clear. All that's needed are indications of how reference seems to be fixed by a set of practices on the part of the linguistic community.

I speculate that considerations of *this* sort are ones that made – at least at first glance – the old paradigm-case argument look plausible. One focused on the practical procedures that we actually use when we refer and overlooked the transcendence of reference – how quickly we are willing to give up these procedures – that eliciting the sceptical set of intuitions turns on.

these two different ways leads to conflicting intuitions because the first version exploits our intuitive allowance for wan mistakes, while the second version exploits our reluctance to accept referential practices too far from our current ones (on pain of making our terms meaningless).

Here's another illustration of how conflicting intuitions due to the split in referential practice between how we speak and what we do give rise to philosophical puzzles. One way that issues about vagueness arise is by observing that our terms seem to be bivalent, without there being ways, even in principle, for deciding borderline cases. Consider the classic *heap* argument. Two items (pebbles, say) do not make a heap; several thousand do. Here's an intuitively plausible principle:

(PH) *Given a heap, removing one item from it doesn't yield a non-heap.*

Unfortunately, stubborn repetition of the above process *does* yield a non-heap. Here's a *solution*. (PH) is wrong. Heaps and non-heaps are crisply individuated (bivalence holds). *But* we don't *know* in every case whether we have a heap or not. And sometimes (in those cases where we're not sure whether we've got a heap), removing *one* pebble does yield a non-heap.

What's *wrong* with this solution?[6] Well, given how we *learn* to pick out heaps and non-heaps, it's hard to see what the purported crisp distinction could consist in (what metaphysical fact could, in a principled way, distinguish between heaps and non-heaps?) that simultaneously could be one *actually embodied in our referential practices*. Let there *be* such a distinction: what in our learning to distinguish heaps and non-heaps corresponds to our language having marked this distinction out?

What's going on? I argued in § 5 that procedures associated with any term do not specify in every circumstance whether or not the items the terms refer to are present; and (this is the epistemic slighting of procedures and the transcendence of reference) we just shunt what our current procedures fail to pick out over to the epistemic side of the equation. This verbal practice is intuitively acceptable when it comes to natural-kind terms (there's always more to learn about gold or viruses), and so the gap between what procedures for these terms pick out and our adoption of bivalence can be treated epistemically without disturbing our intuitive complacency. But what *more* is there to learn about *heaps* that can help us determine for every item whether it's a heap or not? Nothing, and so it looks like we're covering up a metaphysical gap with epistemic wallpaper. (We are, and we're always doing this, but heaps expose the activity in a particularly stark way.)

---

6 In asking this, I mean to ask, "why is this solution so unappealing, *intuitively speaking?*" Otherwise, the solution seems fine.

Heap-style worries can be extended to any term which holds of relatively large objects that can be picked apart into small bits. For, intuitively, we can't see how to justify a refinement or modification of procedures associated with the term to resolve the vague cases, where the bits are there or missing, and so we're tempted to reject the epistemic loosening of procedures we're perfectly willing to accept otherwise.

The intuitions I've described in these illustrations arise from our epistemic idioms failing to mesh intuitively with our referential ones. My aim, I must stress, in discussing these conundra, has not been to solve them, but to show what their intuitive sources are; although in doing this, there is a sense, I think, in which they *have been* solved.

The solution goes something like this. The conflicting intuitions that give rise to the puzzles just discussed are due to our current referential practices. To truly eliminate the sources of these intuitions, we'd have to adopt the language the causal operationalist urged on us; and that's impossible.

Some might feel – especially with regard to heaps – that epistemic notions, "ignorance" in particular, are being used inappropriately. To be ignorant is to be ignorant of *something*, not merely to have nothing to say. To say that there is a border between heaps and non-heaps, but to be ignorant of what it is, requires that – in some sense – *there be a border*. Otherwise, speaking of ignorance in this case is to be irresponsible to the epistemic idioms.

I take this view to be at work in Williamson (1994), and to explain why, when he adopts the ignorance view of vagueness, he feels it is in such need of defense, and why, in particular, he feels required to show there is no successful argument that there is no border between heaps and non-heaps.[7] He postulates (7. 3) an omniscient speaker who, consequently, never says she "doesn't know." As we present her with slowly shrinking heaps (removing grains of sand one by one), we find that she continues to respond "Yes" to the question of whether what's in front of her is a heap, until she falls silent for a while, and then says "No" to the question of whether what's in front of her is a heap. Williamson argues that if we tell omniscient speakers to answer conservatively (p. 200: "so they answer 'yes' to as few questions as is permissible"), we have to expect all omniscient speakers to stop at the same point – otherwise the less conservative ones are not following the instructions.

This won't work because there's little reason to think that "conservative" doesn't admit of vagueness just as "heap" does. But, apart from this, the implicit view about needing to be responsible to the epistemic idiom of

---

7   And yet it's *obvious* no such border exists, in the sense that nothing in what we've learned about how to adjudicate between heaps and non-heaps determines a border. Further, there's likely to be no change in our notion of heaps and non-heaps to cause us to introduce new procedures that provide a crisp border.

ignorance must be rejected: It is only the other side of the coin to the elusivity of *truth*. We are *allowed* to say that *A* is true or *A* is false *independently* of whether our current truth-gathering procedures dictate (even in principle) one or the other; *and* so we are allowed to say we are *ignorant* of which it is. We can say *both* these things without it being the case that – out there in logical space – an adjudication *is* made. So we must allow the expression of ignorance to be admissible even in those cases where it is palpable that there's nothing out there to be ignorant of.[8]

A tradition about the later Wittgenstein I've heard is that he is engaged in philosophy as therapy. One is *cured* of philosophy, and concern with typical philosophical problems, when one sees which misunderstandings about ordinary language it is that they arise from. One learns to see how, in philosophy anyway, language is often on holiday[9] – and it is in recognizing clearly what our practices with certain terms are, that one solves, or dissolves, philosophical problems. From this point of view, ordinary usage is pure and innocent, and getting back to it, and seeing clearly how it operates, cures one of philosophical anguish.[10]

But there's an older and darker tradition, the religious one of original sin, with its accompanying institution of confession.[11] Confession is most naturally accompanied by the more pessimistic view that "talking therapies" ultimately cannot solve psychological problems, because, at root, our problems are not solvable. Rather, the deep problems we have are part of the human condition: they are traps mortality itself sets for us; confession can only soothe (albeit temporarily) our anguish at what we are, and provide small ritualistic acts of absolution as temporary balms.

The latter is analogous to (although more melodramatic than) the perspective I've pressed here. For, first, I've pointed out what it is about our referential practices that causes conflicting philosophical intuitions. Second, I've shown these referential practices are unavoidable. Therefore, third, I've shown we can't escape the intuitions – the solution to the puzzles these intuitions give rise to is our recognition that they're part, as it were, of our linguistic condition. No cure.

---

8   And we do allow this. We're always allowed to say: "I don't know," when faced with a borderline balding case. We don't have to say: "Nothing in my use of the word 'bald' prepared me for anything like *this*."

9   Wittgenstein (1953: § 38).

10 One can also read Kant this way. Reason, unchecked, leads to puzzling paradoxes; keep reason within the bounds of possible experience, and all is cool.

11 Just as there are philosophers who will deny that the Wittgenstein I've described here is anything like the real Wittgenstein, so, too, there are those who will say my description of confession has little to do with the actual practice. Whatever. Regard these as ahistorical parables designed to make clear what I've been trying to do in this book, and how it differs from what other philosophers have tried to do.

There's a deep divide between what we say and what we do. If we look to what we do, there are no problems: we find robust causal relations between tokens uttered and objects scruted – we find a modest (although open-ended) inventory of procedures and dispositions that imply answers to how procedures will operate when applied to new terms.

Similarly, if we look to what we *say*, there are no problems: we find a simple vocabulary, much of which is in the process of being pressed docilely into service by scientific law; we also find a nice logic, and with it a pleasant (and simple) theory of truth and reference.

Moral hypocrites, so the rumor goes, rarely have insomnia. No wonder, if it's true: their ethical accounting system is so much simpler than the one the rest of us have. Naturalizing reference the naive way, that is, bringing our referential *talk* in line with our referential resources (e.g. causality), is the metaphysical attempt to do to our conceptual scheme what the fanatically honest yearn to do to the hypocrite's private life.

Such an attempt, I've argued, exposes two related problems: the projection problem and the apparent transcendence of the referential relation. The gap between what we say and what we do is *indicated* by the transcendence of the reference relation, and is *meant to solve* the pragmatic problems that the projection problem gives rise to. We really do have a good product: we really do connect ourselves causally to the world in subtle ways that teach us a great deal. On the other hand, a good product is good news, and we need a suitable idiom for spreading good news when we have some. The naive view is that a suitable idiom is one that *accurately* portrays what we do. But that's not true.

The point of ethical hypocrisy is comfort, and the story goes that good people sacrifice comfort for truth – where truth is honored either by trimming the talk when it exaggerates or doing whatever more is needed to keep one's word. Here, however, more than mere comfort is at stake. Be overly literal-minded about how we talk about reference, and we lose touch with how much of the world we're actually in contact with. The talk is clean, but the world fades to a construct that impinges on what we say only at the edges, if at all. On the other hand, trim away all talk that either smacks of exaggeration, makes our procedures look cleaner than they are, or elevates the muddy generalization to a golden law, and we virtually lose the capacity to talk at all: we find ourselves swamped in a rich vocabulary that won't extend a whit beyond our fingertips.

The hypocrite, at least the kind of metaphysical hypocrite I urge us to be, finds himself in the best position of all: he can brag with the best of them, and he has inherited the earth, too.

GENERAL CONCLUSION

One way to end a book is gently: to step back and reveal its thematic unity by pursuing, in a general way, threads linking the various topics covered. I do this in what follows, and then discuss a possible tension between the sort of anti-instrumentalism defended in the first part of the book (Parts I and II) with the views of reference developed and defended in the second (Part III, especially).

The failure of various sorts of reductionism links the topics of the elusiveness of reference and truth, on the one hand (Part I § 7, Part III, and Part IV), and the emergence of the special sciences, with their laws and language, and free-floating gross regularities, on the other (Part I).

But the reasons for these failures in the respective cases are not quite the same, although epistemic motives loom in both cases. With the latter the failure is *entirely* due to problems of tractability; and this explains why I defend a kind of physicalism despite the absence of any sort of reducibility in vocabulary (see the end of Part I § 4). In the former the failure is because the idioms in question have special roles: they're *designed*, as it were, to be irreducible.

I showed this in several ways. I considered the naive naturalization program for reference and truth, and showed its incompatibility both with deep-seated intuitions (what we might call intuitions of normativity about these idioms – Part II § 7, Part III § 7, Part IV § 3), as well as pretheoretical constraints on the scope of such idioms (Part IV §§ 1 and 2). I also showed that attempts to replace such idioms by other ones amenable to naturalization are fruitless (Part IV § 4).

Although the irreducibility of the idioms of the special sciences is dissimilar to the irreducibility of the idioms of truth and reference (in particular, because of the normative intuitions present with the latter, but not with the former), one might attempt assimilation anyway, and adopt towards truth and reference the sort of physicalist attitude I urged towards the idioms of the special sciences.

This would be a mistake. The right place to locate the "science" of truth and reference *in general* is in pure mathematics, and as these idioms are used in specific languages (English, for example) in *applied* mathematics.

Causal theories of reference have *masked* this fact about the referential- and truth-idioms because viewpoints compatible with such theories (views taking such idioms as belonging to the special empirical science of seman- tics) take the referential-idiom as explicated in, or reduced to, causal terms. Physicalistically acceptable truth is supposed to follow in turn.

Part III § 7 and Part IV exposed a body of intuitions and practices that make this program implausible. What has gone wrong can be put neatly in terms of an analogy: Suppose someone, noticing the role of geometric objects in physics, propounded the program of reducing *these* to physical ones. Single-mindedly focusing on such a program would cause this person to overlook entirely the very different ways that mathematical points and genuine physical objects, like atoms, operate.

Once we recognize that *reference* and *truth* are pieces of a *mathematical* framework that operates in a way tightly analogous to how *geometry* oper- ates in the study of physical objects, a host of philosophical puzzles subside. No longer need we concern ourselves with how reference and truth (and logic) are to be made compatible with one or other doctrine of physicalism; no longer need we worry about exactly how causality fits with reference, or why various ways of treating reference as a purely causal notion fail; no longer need we wonder why truth plays the roles it plays in explanations of behavior or knowledge. All such questions implicitly treat *truth* and *reference* as non-mathematical idioms – which they simply are not.

In a way, this should be obvious. After all, Tarski gave us a *mathematical* theory. But, the protest goes, one distinguishes a theory of truth from *truth* – from what we take to *be* true. However, *truth* (and *reference*) in this sense is nothing more than an instance of *applied* mathematics, and one burden of this book has been to show that that's exactly how the idiom operates, both intuitively and with regard to its recalcitrance vis-à-vis certain philosophical programs.

Next point. The reader may sense a certain tension between attitudes expressed in the earlier part of the book and attitudes taken later. Here's a way of bringing it out (and resolving it, too). Recall the picture of reference given in Part IV § 5, in particular, the claim that we can unify the language the causal operationalist hoped to saddle us with into something tractable. But "something tractable," after all, is a set of categories that obey nice generalizations. Having gotten this isn't just a bit of amazing luck, is it? Isn't an explanation needed?

Sure, and here's one: we've *really* managed to lock on to the kinds the world is made up of (at least in our neck of the woods). This sounds right at least for some of our generalizations and categories (the most fundamental ones). Aren't we then *referring* to them? Sure, in scare-quotes, anyway, since it doesn't follow that we can causally underwrite reference. On the contrary, for it might be that there's still lots more to *discover* about those kinds, and what's to discover is not implicitly contained in our theories or procedures for handling the kind (we've got a grip on a stuff, say, but *not* on all instances of it). This means that subsequent science can go smoothly, although it doesn't follow that in the fullness of time we'll get (causally) *all* of the stuff in hand.

So, in the first part of the book, I exhibited "realist" proclivities – that is, I urged taking scientific laws and nomenclature seriously, not instrumentally. And this was not simply meant in what some philosophers might regard as a Pickwickian sense – "Well, they're true, and we've got existential quantifiers ranging over the categories, too"; rather, it came out in our noticing that scientific programs are *ongoing*: one *explains* anomalies and mistakes via laws and kinds we have; one applies these things to new situations, and one whittles away at idealizations and checks whether results improve.

But, in the second part of the book, I exhibited "irrealist" tendencies – pointing out how *reference* is something of a put-up job, and explaining, by means of our tendency to cover up the fact that our causal reach is far less than our referential bluff, puzzling intuitions regarding vagueness and scepticism.

Luckily (for me), there is *no* problem reconciling the view about reference in the latter part of the book with the view about science in the first part (and this is especially pleasing since the view in the second half of the book *depends on* that of the first half). The methodological depth that science presupposes simply *doesn't require* referential depth. Our terminology, referentially speaking, invites positing: we "take" our terms to so refer – and *guess* taking them to apply in the way that we talk won't lead to trouble. If we guess right, our science works (as far as we can push it, anyway), and this regardless of whether we ever have the causal resources to underwrite our referential promissory notes.

Putting the matter this way allows me to conclude the book by describing its theme in a unified way: I've been concerned with the role of *causality* in knowledge and reference, where the *causality* pertinent describes the relations we forge and exploit between ourselves and those things in the world we try to talk and learn about.

BIBLIOGRAPHY

Achinstein, Peter (1965) "The Problem of Theoretical Terms," *Readings in the Philosophy of Science*, ed. Baruch A. Brody, (1970), New York: Prentice-Hall, 234–50.

——(1987) "Scientific Discovery and Maxwell's Kinetic Theory," *Particles and Waves* (1991), Oxford: Oxford University Press, 233–57.

——(1990) "The Only Game in Town," *Particles and Waves* (1991), Oxford: Oxford University Press, 259–78.

——(1991) "Theory, Experiment, and Cathode Rays," *Particles and Waves*, Oxford: Oxford University Press, 299–333.

Aczel, Peter (1988) *Lectures on Nonwellfounded Sets*, Stanford, California: CSLI/Stanford University.

Arnold, V.I. (1980) *Ordinary Differential Equations*, Cambridge, MA.: The MIT Press.

Azzouni, Jody (1991) "A Simple Axiomatizable Theory of Truth," *Notre Dame Journal of Formal Logic* 32: 458–93.

——(1994) *Metaphysical Myths, Mathematical Practice: The Ontology and Epistemology of the Exact Sciences*, Cambridge: Cambridge University Press.

——(1997a) "Thick Epistemic Access: Distinguishing the Mathematical from the Empirical," *Journal of Philosophy* 94: 472–84.

——(1997b) "Applied Mathematics, Existential Commitment, and the Quine–Putnam Indispensability Thesis," *Philosophia Mathematica* 5: 193–209.

——(1998) "On 'On What There Is'," *Pacific Philosophical Quarterly* 79: 1–18.

——(forthcoming) "Proof and Ontology in Euclidean Mathematics," *Proceedings of New Trends in the History and Philosophy of Mathematics at Roskilde University*.

Balaguer, Mark (1998) *Platonism and Anti-Platonism in Mathematics*, Oxford: Oxford University Press.

Barwise, J. and Etchemendy, J. (1987) *The Liar: An Essay on Truth and Circularity*, New York: Oxford University Press.

Barwise, J. and Feferman, S. (eds) (1985) *Model-Theoretic Logics*, Berlin: Springer–Verlag.

Basalla, George (1988) *The Evolution of Technology*, Cambridge: Cambridge University Press.

Benacerraf, Paul (1965) "What Numbers Could Not Be," in Paul Benacerraf and Hilary Putnam (eds) *Philosophy of Mathematics* (2nd edn, 1983), Cambridge: Cambridge University Press, 272–94.

Blackburn, Thomas (1988) "The Elusiveness of Reference," in Peter A. French, Theodore E. Uehling, Jr, and Howard K. Wettstein (eds) *Realism and Antirealism: Midwest Studies in Philosophy* 12, Minneapolis, MN: University of Minnesota Press, 179–94.

Block, Ned (ed.) (1980) *Readings in Philosophy of Psychology: Volume 1*, Cambridge, MA: Harvard University Press.

——(1981) *Readings in Philosophy of Psychology: Volume 2*, Cambridge, MA: Harvard University Press.

Boolos, George S. and Jeffrey, Richard C. (1989) *Computability and Logic* (3rd edn), Cambridge: Cambridge University Press.

Bouwsma, O.K. (1949) "Descartes' Evil Genius," *Philosophical Review* 58: 141–51.

Boyd, Richard (1980) "Materialism Without Reductionism: What Physicalism Does Not Entail," in Ned Block (ed.) *Readings in Philosophy of Psychology: Volume 1*, Cambridge, MA: Harvard University Press, 67–106.

——(1981) "Scientific Epistemology and Naturalistic Epistemology," in P.D. Asquith and R.N. Giere (eds) *PSA 1980: Volume 2*, East Lansing, MI: Philosophy of Science Assocation, 613–62.

——(1985) "Lex Orandi Est Lex Credendi," in Paul M. Churchland and Clifford A. Hooker (eds) *Images of Science: Essays on Realism and Empiricism with a Reply from Bas C. van Fraassen*, Chicago: University of Chicago Press, 3–34.

——(1991) "Confirmation, Semantics, and the Interpretation of Scientific Theories," in Richard Boyd, Philip Gasper, and J.D. Trout (eds) *The Philosophy of Science*, Cambridge, MA: The MIT Press, 3–35.

Boyd, Richard, Gasper, Philip and Trout, J.D. (eds) (1993) *The Philosophy of Science*, Cambridge, MA: The MIT Press.

Braun, Martin (1993) *Differential Equations and Their Applications*, New York: Springer–Verlag.

Burge, Tyler (1979) "Individualism and the Mental," in Peter A. French, Theodore E. Uehling, Jr, and Howard K. Wettstein (eds) *Midwest Studies in Philosophy* 4, Minneapolis, MN: University of Minnesota Press, 73–121.

Burkholder, JoAnn M. (1999) "The Lurking Perils of *Pfiesteria*," *Scientific American*, August: 42–9.

Carroll, Lewis (1871) *Alice's Adventures in Wonderland and Through the Looking-Glass* (1960), Harmondsworth, England: Penguin Books.

Cartwright, Nancy (1983) *How the Laws of Physics Lie*, Oxford: Oxford University Press.

Chaisson, Eric J. (1992) "Early Results from the Hubble Space Telescope," *Scientific American*, June: 44–51.

Chang, C.C. and Keisler, Jerome H. (1973) *Model Theory*, Amsterdam: North–Holland.

Churchland, Paul M. (1979) *Scientific Realism and the Plasticity of Mind*, Cambridge: Cambridge University Press.

Collins, H.M. (1985) *Changing Order: Replication and Induction and Scientific Practice* (2nd edn, 1992), Cambridge: Cambridge University Press.

Courant, R. and Hilbert, D. (1937) *Methods of Mathematical Physics, Volume 1*, New York: Interscience Publishers, Inc.

——(1962) *Methods of Mathematical Physics, Volume 2*, New York: Interscience Publishers, Inc.

Davidson, Donald (1963) "Actions, Reasons, and Causes," *Essays on Action and Events*, (1980) Oxford: Oxford University Press, 5–19.

——(1973) "Radical Interpretation," *Inquiries into Truth and Interpretation* (1980), Oxford: Clarendon Press, 125–39.

——(1977) "Reality Without Reference," *Inquiries into Truth and Interpretation*, (1980), Oxford: Clarendon Press, 215–25.

——(1979) "The Inscrutability of Reference," *Inquiries into Truth and Interpretation* (1980), Oxford: Clarendon Press, 227–41.

Dawkins, Richard (1987) *The Blind Watchmaker*, New York: W.W. Norton & Company.

Descartes, René (1641) *Meditations on First Philosophy*, in *The Philosophical Works of Descartes* (1931), trans. Elizabeth S. Haldane and G.R.T. Ross, Cambridge: Cambridge University Press, 131–99.

Devitt, Michael (1983) "Realism and the Renegade Putnam: A Critical Study of 'Meaning and the Moral Sciences,'" *Noûs* 17: 291–301.

——(1991) *Realism and Truth* (2nd edn), Oxford: Blackwell.

Dreben, Burton and Goldfarb, Warren D. (1979) *The Decision Problem*, Reading, MA: Addison-Wesley.

Dretske, Fred I. (1981) *Knowledge and the Flow of Information*, Cambridge, MA: The MIT Press.

Duhem, Pierre (1914) *The Aim and Structure of Physical Theory* (1962), trans. Philip P. Wiener, New York: Atheneum.

——(1969) *To Save the Phenomena*, trans. Edmund Doland and Chanenah Maschler, Chicago: University of Chicago Press.

Dupré, John (1981) "Natural Kinds and Biological Taxa," *Philosophical Review* 90: 66–90.

——(1993) *The Disorder of Things*, Cambridge, MA: Harvard University Press.

Eiceman, G.A., Leasure, C.S., and Vandiver, V.J. (1986) "Negative Ion Mobility Spectrometry for Selected Inorganic Pollutant Gases and Gas Mixtures in Air," *Analytical Chemistry* 58: 76–80.

Ellis, Brian (1985) "What Science Aims to Do," in Paul M. Churchland and C. W. Hooker (eds) *Images of Science: Essays on Realism and Empiricism, With a Reply from Bas C. van Fraassen*, Chicago: University of Chicago Press, 48–74.

Enderton, Herbert B. (1972) *A Mathematical Introduction to Logic*, New York: Academic Press.

Evans, Gareth (1973) "The Causal Theory of Names," *Collected Papers* (1985), Oxford: Oxford University Press, 1–24.

Feyerabend, Paul K. (1962) "Explanation, Reduction and Empiricism," *Realism, Rationalism and Scientific Method: Volume 1* (1981), London: Cambridge University Press, 44–96.

Feynman, Richard P., Leighton, Robert B., and Sands, Matthew (1963) *The Feynman Lectures on Physics*, Reading, MA: Addison-Wesley.

Field, Hartry (1972) "Tarski's Theory of Truth," *Journal of Philosophy* 64: 347–75.

——(1975) "Conventionalism and Instrumentalism in Semantics," *Noûs* 9: 376–406.

——(1981) "Mental Representation," in Ned Block (ed.) *Readings in Philosophy of Psychology, Volume 2*, Cambridge, MA: Harvard University Press, 78–114.

——(1989) *Realism, Mathematics and Modality*, Oxford: Blackwell.

——(1996) "The Aprioricity of Logic," *Proceedings of the Aristotelian Society* 96: 358–79.

Fodor, Jerry A. (1974) "Special Sciences," *Representations* (1981), Cambridge, MA: The MIT Press, 127–45.

——(1979) *The Language of Thought*, Cambridge, MA: Harvard University Press.

——(1983) *The Modularity of Mind*, Cambridge, MA: The MIT Press.

——(1987) *Psychosemantics*, Cambridge, MA: The MIT Press.

——(1990a) "Stephen Schiffer's Dark Night of the Soul: A Review of *Remnants of Meaning*," *A Theory of Content and Other Essays*, Cambridge, MA: The MIT Press, 177–91.

——(1990b) "A Theory of Content I: The Problem," *A Theory of Content and Other Essays*, Cambridge, MA: The MIT Press, 51–87.

——(1990c) "A Theory of Content II: The Theory," *A Theory of Content and Other Essays*, Cambridge, MA: The MIT Press, 88–136.

Franklin, Allan (1993) *The Rise and Fall of the Fifth Force*, New York: American Institute of Physics.

Garfinkel, Alan (1981) *Forms of Explanation*, New Haven, CT: Yale University Press.

Giere, Ronald N. (1985) "Constructive Realism," in Paul M. Churchland and C. W. Hooker (eds) *Images of Science: Essays on Realism and Empiricism, With a Reply from Bas C. van Fraassen*, Chicago: University of Chicago Press, 75–98.

Glymour, Clark (1982) "Conceptual Scheming, or Confessions of a Metaphysical Realist," *Synthese* 51: 169–80.

Goffman, Erving (1971) "Normal Appearances," *Relations in Public*, New York: Harper Colophon Books, 238–333.

Goldman, Alvin (1986) *Epistemology and Cognition*, Cambridge, MA: Harvard University Press.

Goodman, Nelson (1955) *Fact, Fiction, and Forecast* (4th edn, 1983), Cambridge, MA: Harvard University Press.

Gould, Stephen (1981) *The Mismeasure of Man*, New York: W.W. Norton & Company.

Grünbaum, Adolf (1954) "Science and Ideology," *The Scientific Monthly*, July: 13–19.

Hacking, Ian (1981) "Do We See Through a Microscope?," in Paul M. Churchland and C. A. Hooker (eds) *Images of Science: Essays on Realism and Empiricism, With a Reply from Bas C. van Fraassen*, (1985), Chicago: University of Chicago Press, 132–52.

——(1983) *Representing and Intervening*, London: Cambridge University Press.

——(1990) *The Taming of Chance*, London: Cambridge University Press.

Haj-Hussein, Amin T., Christian, Gary D., and Ruzicka, Jaromir (1986) "Determination of Cyanide by Atomic Absorption Using a Flow Injection Conversion Method," *Analytical Chemistry* 58: 38–42.

Hardin, C.L. (1988) *Color for Philosophers*, Indianapolis, IN: Hackett Publishing Company.

Harman, Gilbert (1965) "The Inference to the Best Explanation," *Philosophical Review* 74: 88–95.

Hart, H.L.A. and Honoré, A.M. (1959) *Causation in the Law*, Oxford: Oxford University Press.

Hatfield, Gary (1990) "Metaphysics and the New Science," in David C. Lindberg and Robert S. Westman (eds) *Reappraisals of the Scientific Revolution*, Cambridge: Cambridge University Press, 93–166.

Heck, Richard G., Jr (1997) "Tarski, Truth, and Semantics," *The Philosophical Review* 106: 533–54.

Hempel, Carl G. (1954) "A Logical Appraisal of Operationism," *Aspects of Scientific Explanation*, New York: The Free Press, 123–33.

——(1988) "Provisos: A Problem Concerning the Inferential Function of Scientific Theories," *Erkenntnis* 28: 147–64.

Horwich, Paul (1990) *Truth*, Oxford: Blackwell (reissued (1998) by Oxford University Press).

Hoskin, Michael (1995) "The Discovery of Uranus, the Titius–Bode Law, and the Asteroids," in René Taton and Curtis Wilson (eds) *Planetary Astronomy from the Renaissance to the Rise of Astrophysics*, Part B: *The Eighteenth and Nineteenth Centuries*, Cambridge: Cambridge University Press, 169–80.

Kaku, Michio (1993) *Quantum Field Theory*, Oxford: Oxford University Press.

Kateman, G. and Buydens, L. (1993) *Quality Control in Analytical Chemistry* (2nd edn), New York: John Wiley & Sons, Inc.

Kauffman, Stuart A. (1993) *The Origins of Order*, Oxford: Oxford University Press.

Kearns, Edward, Kajita, Takaaki, and Totsuka, Yoji (1999) "Detecting Massive Neutrinos," *Scientific American*, August: 64–71.

Kim, Jaegwon (1988) "What Is 'Naturalized Epistemology'?," *Supervenience and Mind: Selected Philosophical Essays* (1993), Cambridge: Cambridge University Press, 216–36.

——(1989) "The Myth of Nonreductive Materialism," *Supervenience and Mind: Selected Philosophical Essays* (1993), Cambridge: Cambridge University Press, 265–84.

——(1992) "Multiple Realization and the Metaphysics of Reduction," *Supervenience and Mind: Selected Philosophical Essays* (1993), Cambridge: Cambridge University Press, 309–35.

Kindel, Stephen (1992) "Styling and Design: Chrysler," *Financial World*, April 14: 42.

Kitcher, Philip (1984) "1953 and All That: A Tale of Two Sciences," *The Philosophical Review* 93: 335–73. *The Philosophy of Science*, eds Richard Boyd, Philip Gasper, J.D. Trout, Cambridge, MA: The MIT Press, 553–70.

Kline, Morris (1972) *Mathematical Thought from Ancient to Modern Times*, Oxford: Oxford University Press.

Kornblith, Hilary (ed.) (1985) *Naturalizing Epistemology*, Cambridge, MA: The MIT Press.

Kripke, Saul (1979) "Speaker's Reference and Semantic Reference," in Gary Ostertag (ed.) *Definite Descriptions: A Reader*, Cambridge, MA: The MIT Press, 225–56.

——(1980) *Naming and Necessity*, Cambridge, MA: Harvard University Press.

Kuhn, Thomas S. (1961) "The Function of Measurement in Modern Physical Science," *The Essential Tension* (1977), Chicago: University of Chicago Press, 178–224.

——(1970) *The Structure of Scientific Revolutions*, Chicago: University of Chicago Press.

——(1977) "Objectivity, Value Judgment, and Theory Choice," *The Essential Tension*, Chicago: University of Chicago Press, 320–39.

Lakatos, Imre (1976) "Falsification and the Methodology of Scientific Research Programmes," in Imre Lakatos and Alan Musgrave (eds) *Criticism and the Growth of Knowledge*, London: Cambridge University Press, 91–196.

Leeds, Stephen (1978) "Theories of Reference and Truth," *Erkenntnis* 13: 111–29.

Lewis, David (1983) "New Work for a Theory of Universals," *Papers in Metaphysics and Epistemology* (1999), Cambridge: Cambridge University Press, 8–55.

——(1984) "Putnam's Paradox," *Papers in Metaphysics and Epistemology* (1999), Cambridge: Cambridge University Press, 56–77.

Lewis, Harry R. (1979) *Unsolvable Classes of Quantificational Formulas*, Reading, MA: Addison-Wesley.

Maxwell, Grover (1962) "The Ontological Status of Theoretical Entities," in Baruch A. Brody (ed.) *Readings in the Philosophy of Science*, New York: Prentice-Hall, 224–33.

Maxwell, James Clerk (1965) "On the Dynamical Evidence of the Molecular Constitution of Bodies," *The Scientific Papers of James Clerk Maxwell*, ed. W. D. Niven, New York: Dover Publications, 418–38.

McGee, Vann (1991) *Truth, Vagueness, and Paradox*, Indianapolis, IN: Hackett Publishing Company.

Merrill, G.H. (1980) "The Model-Theoretic Argument Against Realism," *Philosophy of Science* 47: 69–81.

Millikan, Ruth Garrett (1984) *Language, Thought, and Other Biological Categories*, Cambridge, MA: The MIT Press.

Nagel, Ernest (1961) *The Structure of Science*, New York: Harcourt, Brace & World Inc.

Neher, Erwin and Sakmann, Bert (1992) "The Patch Clamp Technique," *Scientific American*, March: 44–51.

Polanyi, Michael (1958) *Personal Knowledge: Towards a Post-Critical Philosophy*, New York: Harper Torchbooks.

Popper, Karl (1974) "Autobiography of Karl Popper," *The Philosophy of Karl Popper*, ed. Paul Arthur Schilpp, La Salle, IL: Open Court, 3–181.

Powell, Corey S. (1992) "The Golden Age of Cosmology," *Scientific American*, July: 17–22.

Purcell, Edward M. (1985) *Electricity and Magnetism (Berkeley Physics Course): Volume 2*, New York: McGraw-Hill, Inc.

Putnam, Hilary (1970) "Is Semantics Possible?," *Mind, Language and Reality: Philosophical Papers, Volume 2* (1975), Cambridge: Cambridge University Press, 139–52.

——(1975a) "The Meaning of 'Meaning'," *Mind, Language and Reality: Philosophical Papers, Volume 2*, Cambridge: Cambridge University Press, 215–71.

——(1975b) "Philosophy and Our Mental Life," *Mind, Language and Reality: Philosophical Papers, Volume 2*, Cambridge: Cambridge University Press, 291–303.

——(1977) "Models and Reality," *Realism and Reason: Philosophical Papers, Volume 3*, (1983), Cambridge: Cambridge University Press, 1–25.

——(1978a) *Meaning and the Moral Sciences*, London: Routledge & Kegan Paul.

——(1978b) "Realism and Reason," *Meaning and the Moral Sciences*, London: Routledge & Kegan Paul, 123–40.

——(1978c) "Reference and Understanding," *Meaning and the Moral Sciences*, London: Routledge & Kegan Paul, 97–119.

——(1981) *Reason, Truth and History*, Cambridge: Cambridge University Press.

——(1983a) "Beyond Historicism," *Realism and Reason: Philosophical Papers, Volume 3*, Cambridge: Cambridge University Press, 287–303.

——(1983b) "Introduction: An Overview of the Problem," *Realism and Reason: Philosophical Papers, Volume 3*, Cambridge: Cambridge University Press, vii–xviii.

——(1983c) "Vagueness and Alternative Logic," *Realism and Reason: Philosophical Papers, Volume 3*, Cambridge: Cambridge University Press, 271–86.

——(1983d) "Why Reason Can't Be Naturalized," *Realism and Reason: Philosophical Papers, Volume 3*, Cambridge: Cambridge University Press, 229–47.

——(1983e) "Why There Isn't a Ready-Made World," *Realism and Reason: Philosophical Papers, Volume 3*, Cambridge: Cambridge University Press, 205–28.

——(1984) "Is the Causal Structure of the Physical Itself Something Physical?," in Peter A. French, Theodore E. Uehling, Jr, and Howard K. Wettstein (eds) *Midwest Studies in Philosophy* 9, Minneapolis: University of Minnesota Press, 3–16.

——(1986) "Information and the Mental," in Ernest Lepore (ed.) *Truth and Interpretation: Perspectives on the Philosophy of Donald Davidson*, Oxford: Blackwell, 362–27.

——(1987) *The Many Faces of Realism*, LaSalle, IL: Open Court.

——(1989) "Model Theory and the 'Factuality' of Semantics," in Alexander George (ed.) *Reflections on Chomsky*, Oxford: Blackwell, 213–32.

——(1992) *Renewing Philosophy*, Cambridge, MA: Harvard University Press.

Quine, W. V. (1935) "Truth by Convention," *The Ways of Paradox and Other Essays* (rev. enl. edn, 1976), Cambridge, MA: Harvard University Press, 77–106.

——(1951) "On What There Is," *From a Logical Point of View* (1980) Cambridge, MA: Harvard University Press, 1–19.

——(1960) *Word and Object*, Cambridge, MA: The MIT Press.

——(1964) "Ontological Reduction and the World of Numbers," *The Ways of Paradox and Other Essays* (rev. enl. edn, 1976), Cambridge, MA: Harvard University Press, 212–20.

——(1969a) "Epistemology Naturalized," *Ontological Relativity and Other Essays*, New York: Columbia University Press, 69–90.

——(1969b) "Natural Kinds," *Ontological Relativity and Other Essays*, New York: Columbia University Press, 114–38.

——(1969c) "Ontological Relativity," *Ontological Relativity and Other Essays*, New York: Columbia University Press, 26–68.

——(1969d) *Set Theory and its Logic* (rev. edn), Cambridge, MA: Harvard University Press.

——(1970a) "On the Reasons for Indeterminacy of Translation," *Journal of Philosophy* 67: 178–83.

——(1970b) *Philosophy of Logic*, Englewood Cliffs, NJ: Prentice-Hall.

——(1975) "The Nature of Natural Knowledge," in Samuel Guttenplan (ed.) *Mind and Language*, Oxford: Oxford University Press, 67–81.

——(1980) "Foreword, 1980," *From a Logical Point of View*, Cambridge, MA: Harvard University Press, vii–xii.

——(1981a) "Reply to Stroud," in Peter A. French, Theodore E. Uehling, Jr, Howard K. Wettstein (eds) *Midwest Studies in Philosophy* 6, Minneapolis: University of Minnesota Press, 473–75.

——(1981b) "Responses," *Theories and Things*, Cambridge, MA: Harvard University Press, 173–86.

——(1981c) "Things and Their Place in Theories," *Theories and Things*, Cambridge, MA: Harvard University Press, 1–23.

——(1981d) "What Price Bivalence?," *Theories and Things*, Cambridge, MA: Harvard University Press, 31–7.

——(1986a) "Reply to Henryk Skolimowski," in Lewis Edwin Hahn and Paul Arthur Schilpp (eds) *The Philosophy of W.V. Quine*, La Salle, IL: Open Court, 492–93.

——(1986b) "Reply to Paul A. Roth," in Lewis Edwin Hahn and Paul Arthur Schilpp (eds) *The Philosophy of W.V. Quine*, La Salle, IL: Open Court, 459–61.

——(1987) *Quiddities: An Intermittently Philosophical Dictionary*, Cambridge, MA: Harvard University Press.

Radnitzky, Gerard and Bartley, W.W., III. (eds) (1987) *Evolutionary Epistemology, the Theory of Rationality, and the Sociology of Knowledge*, La Salle, IL: Open Court.

Rapsomanikis, Spyridon, Donard, O.F.X., and Weber, James, H. (1986) "Speciation of Lead and Methyllead Ions in Water by Chromatography/Atomic Absorption Spectrometry after Ethylation with Sodium Tetraethylborate," *Analytical Chemistry* 58: 35–8.

Reed, Michael and Simon, Barry (1978) *Methods of Modern Mathematical Physics*, Volume 4: *Analysis of Operators*, New York: Academic Press.

Resnick, Robert, Halliday, David, and Krane, Kenneth S. (1992) *Physics*, 2 vols (4th edn), New York: John Wiley & Sons, Inc.

Resnik, Michael D. (1989) "Computation and Mathematical Empiricism," *Philosophical Topics* 17: 129–44.

——(1990) "Immanent Truth," *Mind* 99: 405–24.

——(1997) *Mathematics as a Science of Patterns*, Oxford: Oxford University Press.

Richard, Mark (1997) "Inscrutability," *Canadian Journal of Philosophy* 23: (supplementary volume) 165–209.

Rorty, Richard (1986) "Pragmatism, Davidson and Truth," *Objectivity, Relativism, and Truth: Philosophical Papers, Volume 1* (1991), Cambridge: Cambridge University Press, 126–50.

——(1993) "Putnam and the Relativist Menace," *Journal of Philosophy* 90: 443–61.

Rosenberg, Alexander (1985) *The Structure of Biological Science*, Cambridge: Cambridge University Press.

Rosenthal, David M. (ed.) (1971) *Materialism and the Mind–Body Problem*, Englewood Cliffs, NJ: Prentice-Hall.

—— (ed.) (1991), *The Nature of Mind*, Oxford: Oxford University Press.

Rouchaud, Jean-Claude and Fedoroff, Michel (1986) "Determination of Nitrogen in Metals and Semiconductors by Thermal Neutron Activation," *Analytical Chemistry* 58: 108–09.

Rummelhart, David E., McClelland, James L., and the PDP Research Group (eds) (1986) *Parallel Distributed Processing: Explorations in the Microstructure of Cognition*, Volume 1: *Foundations*, Cambridge, MA: The MIT Press.

Sánchez, F. García and Blanco, C. Cruces (1986) "Determination of the Carbamate Herbicide Propham by Synchronous Derivative Spectrofluorometry following Fluorescamine Fluorogenic Labeling," *Analytical Chemistry* 58: 73–6.

Saunders, B.A.C. and Brakel, J. van (1988) "Re-evaluating Basic Colour Terms," *Cultural Dynamics* 1: 359–78.

Schwartz, Stephen P. (ed.) (1977) *Naming, Necessity, and Natural Kinds*, Ithaca, NY: Cornell University Press.

Searle, John (1983) *Intentionality: An Essay in the Philosophy of Mind*, Cambridge: Cambridge University Press.

——(1984) *Minds, Brains, and Science*, Cambridge, MA: Harvard University Press.

Shoenfield, Joseph R. (1967) *Mathematical Logic*, Reading, MA: Addison-Wesley.

Smith, George (forthcoming) "From the Phenomenon of the Ellipse to an Inverse-Square Force: Why Not?" *Steinfest: In honor of Howard Stein's 70th Birthday*, La Salle, IL: Open Court Publishing.

Smith, George and Mindell, David A. (forthcoming) "The Emergence of the Turbofan Engine," Peter Galison and Alex Roland (eds) *Archimedes*.

Smolensky, P. (1986) "Information Processing in Dynamical Systems: Foundations of Harmony Theory," in David E. Rummelhart, James L. McClelland, and the PDP Research Group (eds) *Parallel Distributed Processing: Explorations in the Microstructure of Cognition*, Volume 1: *Foundations*, Cambridge, MA: The MIT Press, 194–281.

Snyder, Solomon H. and Bredt, David S. (1992) "Biological Roles of Nitric Oxide," *Scientific American*, May: 68–77.

Sosa, Ernest (1993) "Putnam's Pragmatic Realism," *Journal of Philosophy* 90: 605–26.

Stalnaker, Robert C. (1987) *Inquiry*, Cambridge, MA: The MIT Press.

Stoker, Stephen H. (1990) *Introduction to Chemical Principles* (3rd edn), New York: Macmillan Publishing Company.

Stoytcheva, Margarita (1992) "Bioelectrocatalytical Method for Glucose Determination in Foodstuffs," *Bull. Soc. Chim. Belg.* 101: 1043–6.

Strasberg, M. (1991) "Acoustical Measurements," in Rita G. Lerner and George L. Trigg (eds) *Encyclopedia of Physics*, New York: VCH Publishers, 9–13.

Strobel, Howard A. and Heineman, William R. (1989) *Chemical Instrumentation: A Systematic Approach* (3rd edn), New York: John Wiley & Sons.

Stroud, Barry (1981) "The Significance of Naturalized Epistemology," in Peter A. French, Theodore E. Uehling, Jr, and Howard K. Wettstein (eds) *Midwest Studies in Philosophy* 6, Minneapolis: University of Minnesota Press, 455–71.

——(1984) *The Significance of Philosophical Scepticism*, Oxford: Oxford University Press.

Suppe, Frederick (1989a) "Theoretical Perspectives on Closure," *The Semantic Conception of Theories and Scientific Realism*, Chicago: University of Illinois Press, 278–96.

——(1989b) "What's Wrong with the Received View on the Structure of Scientific Theories?" *The Semantic Conception of Theories and Scientific Realism*, Chicago: University of Illinois Press, 38–77.

Suppes, Patrick (1984) *Probabilistic Metaphysics*, Oxford: Blackwell.

Sussman, Gerald Jay and Wisdom, Jack (1992) "Chaotic Evolution of the Solar System," *Science* 257: 56–62.

Tarski, Alfred (1944) "The Semantic Conception of Truth," *Philosophy and Phenomenological Research* 4: 341–75.
——(1956) "The Concept of Truth in Formalized Languages," in John Corcoran (ed.) and J.H. Woodger (trans.) *Logic, Semantics, Metamathematics* (2nd edn, 1983), Indianapolis, IN: Hackett, 152–278.
Thomas, Sir John Meurig (1992) "Solid Acid Catalysts," *Scientific American*, April: 112–18.
Unger, Peter (1983) "The Causal Theory of Reference," *Philosophical Studies* 43: 1–45.
Urquhart, Alasdair (1986) "Many-Valued Logics," in D. Gabbay and F. Guenthner (eds) *Handbook of Philosophical Logic, Volume III*, Dordrecht: D. Reidel, 71–116.
Van Benthem, Johan and Doets, Kees (1983) "Higher-Order Logic," in D. Gabbay and F. Guenthner (eds) *Handbook of Philosophical Logic, Volume I*, Dordrecht: D. Reidel, 275–330.
Van Brakel, J. (1993) "The Plasticity of Categories: The Case of Colour," *British Journal of the Philosophy of Science* 44: 103–35.
Van Fraassen, Bas C. (1980) *The Scientific Image*, Oxford: Oxford University Press.
Varney, R.N. (1991) "Kinetic Theory," in Rita G. Lerner and George L. Trigg (eds) *Encyclopedia of Physics*, New York: VCH Publishers, 601–5.
Wallace, John (1979) "Only in the Context of a Sentence Do Words Have Any Meaning," in Peter A. French, Theodore E. Uehling, Jr, and Howard K. Wettstein (eds) *Contemporary Perspectives in the Philosophy of Language*, Minneapolis: University of Minnesota Press, 305–25 (rev. enl. edn of *Midwest Studies in Philosophy*, Volume 2: *Studies in Philosophy of Language*, 1977).
Weiskrantz, L. (1986) *Blindsight: A Case Study and Implications*, Oxford: Oxford University Press.
Weisman, Richard (1984) *Witchcraft, Magic, and Religion in Seventeenth-Century Massachusetts*, Amherst, MA: Harvard University Press.
Williams, Michael (1980) "Coherence, Justification and Truth," *Review of Metaphysics* 34: 243–72.
Williamson, Timothy (1994) *Vagueness*, London: Routledge.
Wilson, Curtis (1972) "How Did Kepler Discover His First Two Laws?," *Scientific American* 226 (3): 93–106.
——(1980) "Perturbations and Solar Tables from Lacaille to Delambre: The Rapprochement of Observation and Theory, Part I," *Archive for History of Exact Sciences* 22.
Wilson, Mark (1982) "Predicate Meets Property," *Philosophical Review* 41: 549–89.
Wirsz, Douglas F. and Blades, M.W. (1986) "Application of Pattern Recognition and Factor Analysis to Inductively Coupled Plasma Optical Emission Spectra," *Analytical Chemistry* 58: 51–7.
Wittgenstein, Ludwig (1953) *Philosophical Investigations* (1968), trans. G. E. M. Anscombe, New York: Macmillan.
Wright, Crispin (1992) *Truth and Objectivity*, Cambridge, MA: Harvard University Press.
Xingguo, Chen, Zhide, Huo, and Hongwen, Xheng (1992) "Catalytic Spectrophotometric Determination of Ruthenium by Flow Injection Analysis Using Amido Black 10B and KIO_4," *Bull. Soc. Chim. Belg.* 101: 989–94.

INDEX

abstracta 16n, 50, 51–2, 176; problem for naturalism 7–9
Achinstein, Peter 3, 22n, 23n, 35–6, 41n, 216n
Aczel, Peter 173n
Alice's rejoinder 147, 159, 161–2
anti-realism 242; scientific 14, 48
Archimedes 71
Aristotle 180, 189
Arnold, V.I. 19n

Baconian facts *see* observational regularities
Balaguer, Mark 52n, 53n, 89
Bartley, W.W. 4
Barwise, Jon 144n, 173n
Basalla, George 67n-8n, 70
Benacerraf, Paul 93n
biotechnical eyeballs 79–80, 120
bivalence 225, 236–7
Blackburn, Thomas 4n, 172n, 231
Blades, M.W. 72n
Blanco, C. Cruces 72n
blind-sight 175–6
Boolos, George 144n
boundary layer between theory and data 33, 36, 37, 63, 70, 71
Boutroux, Emile 14n
Bouwsma, O.K. 235n
Boyd, Richard 3n, 4n, 56, 60, 136n, 141n, 216n
Boyd, T.A. 65
Boyle's law 43n
Braun, Martin 18n
Bredt, David S. 65n, 69n
Burge, Tyler 206n
Burkholder, JoAnn 153n

Buydens, L. 215n

Carnap, Rudolf 13, 87
Carroll, Lewis 147n; *see also* Alice's rejoinder
Cartesian grid 147
Cartwright, Nancy 14, 33n, 25n, 27, 29–30, 48–9, 52, 54–6, 59n, 61, 62
causal operationalism 190, 215–21, 222, 225, 233, 237
causal scepticism *see* Putnam's "sceptical strategy"
causality 4–6, 71, 72, 148–56, 158–9, 232, 242; interest relativity of 173–4; observability of 170–2; relevance to individuation 152–7; scientific respectability of 174–6
Chaisson, Eric J. 80n
Chang, C.C. 144n, 145n
charity, Davidsonian 231
chemical instrumentation 72n
chemical stoichiometry 28, 34n
Chomsky, Noam 33
Churchland, Paul 15, 74n, 104n, 121
classical operationalism *see* operationalism
COBE satellite 175
coherence, as a matter of application 136
Collins, H.P. 70, 110n, 112n, 137n
commonsense regularities 69; *see also* gross regularities
conceptual change 15, 89; *see also* revisability; unrevisability
conceptual relativity 167n
conceptual scheme, definition of 90
conferva rivularis 108n
confession, as an analogy 238